Hygiologic-Physiscopic Upper Trunk Carapace Tachymetry in Customary and Anemal Phthisis Aegrotants Using Thoracogram

Negahban Zawadi

Hygiologic-Physiscopic Upper Trunk Carapace Tachymetry in Customary and Anemal Phthisis Aegrotants Using Thoracogram

First Edition August 2025

Written by Negahban Zawadi

CONTENTS

INTRODUCTION

The upper part of the trunk is referred to as the thorax. It consists of a thoracic wall which itself include soft tissues and musculature of and an external musculoskeletal cage and some of the internal organs such as lungs, heart, thymus, trachea, esophagus, vagus and phrenic nerves lie in an internal thoracic cavity with the sympathetic trunks (right and left), thoracic duct and systemic and pulmonary blood vessels. Inferiorly diaphragm separates the thorax from the abdominal cavity and superiorly it interconnects with the neck and the upper limbs. The thoracic wall also confers shield to the abdominal viscera. The major part of the liver resides below the right dome of diaphragm; while the left dome comprises the spleen and stomach. The posterior aspects of the superior pole of the kidneys lie on the diaphragm and is anterior to the twelfth rib on the right side. While posterior pole of left kidney lies anterior to the eleventh rib on the left dome of diaphragm posteriorly on the left side.

Variability in thoracic cage dimensions is present partially personal in various people, but they are also related to gender, age of the individual, and the human races. It has been found that a newborn child's transverse diameter is much lesser in comparison to adults. The proportion of a person's transverse diameter grows in size as they learn to walk. In terms of absolute and proportional measures, females are depicted to have a smaller thoracic measurement than males. Females tend to have a shorter vertical expanse of the sternum as compared to the suprasternal notch to the xiphoid cartilage, which poses the inlet of the thoracic cavity to be more inclined. Females have been found to have the jugular notch in level with the third thoracic vertebra posteriorly [1]. Males tend to have a longer sternum in comparison to females, consequently, the thoracic inlet is less sloped and the suprasternal notch is present at the same level as the second thoracic vertebra at the posterior thoracic wall [1]. The size of the thoracic cage diverges between individual's dependent rib mobility and diaphragm expansion during breathing, as well as the distension degree of abdominal viscera.

Human embryos have been found to lack a thorax till they are supposed to reach the final stage of the embryonic development, which usually occurs at the end of the eighth week of intrauterine life[2]. The organogenesis phage is found to be active between the third to eighth weeks of intrauterine development[2]. Thorax is formed around the primitive pericardial cavity and the pericardioperitoneal canals that run through it. The pericardioperitoneal canals and the region of the peritoneal cavity around the lower end of the foregut that grows into the esophagus, the stomach, and the duodenum reside within the cage of the thorax, establishing the pleural cavities that surround these lungs. The formation of the diaphragm additionally precincts these peritoneal cavity sections into the abdominal cavity.

At the 14th stage of the embryonic development, the heart which grows is located at the level of upper cervical somites and above upper limb buds[3]. The hypothetical thoracic zone is formed from the somites of the thorax on the contrary side of the midgut, the pericardial cavity ventrally, and the pericardioperitoneal canals posteriorly, on each side of the foregut[3]. The pleural chambers of the pericardioperitoneal canals have yet to be explored. Endodermal epithelial cells from the foregut, hepatic primordium wall are invaginating septum transversum mesenchyme from the caudal pericardial wall below the heart. The diaphragm which is the lowest part of the thorax might be seen in embryos as early as stage 13[3], and it drifts caudally with the development of craniocaudal and elongation of the neck. The parietal pericardium remains to be connected with the diaphragm as it lowers[3].

The respiratory diverticulum develops as an extension from the front wall of the foregut when an embryo is of approx. 28 days, predisposed by retinoic acid [4] produced by adjacent mesoderm. The transcription factor TBX4 is seen to be upregulated as the concentration of retinoic acid is augmented. This TBX4 promotes lung bud formation along with lung growth and its differentiation[4]. Splanchnopleuric mesenchyme is formed out of the pericardioperitoneal canals' medial walls and thereby protects lung buds, while somatopleuric mesenchyme is a derivative of the lateral walls and adds up to the body wall[3]. This later mesenchyme is seen to be penetrated by the developing ribs which tend to originate from the thoracic sclerotomes. The somatopleuric mesenchyme grounds the sternum and adjoining costal cartilages to rise in the midline[3]. The somatopleuric mesenchyme in the midline creates the sternum and the costal cartilage. Intercostal muscles develop from the ventral edge of the somites' epithelial plate and pass into the bony and cartilaginous enclosure. After the somatopleuric coelomic epithelium proliferates, the parietal layer mesothelium of the pleura develops.

The pulmonary ridge develops on the lateral wall of the pleural coelom and moderately borders the pleuropericardial canal. The ridge is seen to run parallel to the dorsolateral edge of the septum transversum. The ridge and the developing lung bud are firmly associated, splitting it into two membranes that shelter the septum transversum. The characteristic cardinal vein and phrenic nerve are rooted towards the inner portion of the pleuropericardial membrane and travel cranially through the pleuropericardial membrane before meeting the septum transversum[5]. The other membrane that instigates caudally is the pleuroperitoneal membrane. As the lung starts to expand, the apical section inhabits and breaches the body wall, spreading cranially on the lateral side of the shared cardinal vein, tailed by an extension from the main pleural coelom to form part of the secondary, definitive pleural sac. In this alignment, the phrenic nerve and the common cardinal vein are located medially in the mediastinum. The pleuropericardial canal, which runs parallel to the vein and narrows to a slit, hastily fades due to the apposition and fusing of its borders. Closure happens early and is frequently dependent on the development and extension of the thoracic cavity nearby the viscera, which includes the heart and major trachea, esophagus arteries, and lungs rather than the active growth of the pleuropericardial membrane over the aperture[5].

The lung, and its associated visceral and parietal pleurae, raise ventromedially and caudodorsally additional to the cranially. With the ventromedial extension, the lungs and pleurae exhume and split the somatopleuric mesenchyme over the pericardium, splitting it from the ventral and lateral thoracic walls. The ventrolateral fibrous pericardium, parietal serous pericardium, and mediastinal parietal pleura are all somatopleuric in the cause, notwithstanding their topographic width.

The thoracic cage entails typically 12 pairs of ribs, 12 thoracic vertebrae, and a single-segmented sternum in healthy subjects. The tubercle characteristically articulates with the transverse process at costotransverse joints, and rib heads posteriorly articulate with the consistent vertebral bodies with the head at the costovertebral joint. Only the 1st to 7th ribs has vertebrosternal ribs, which have straight costal cartilages that link with the sternum anteriorly. The 8th, 9th, and 10th ribs articulate with the vertebral column posteriorly but do not essentially contact the sternum anteriorly; instead, their costal cartilages are composed collectively and subsequently connect with the 7th rib's costal cartilage.

Looking back to the ancient anatomical literature for identification of the average normal number of ribs, it seems that Flemish Anatomist Andreas Vessalius was one of the first to write regarding the number of ribs in his specialized book of Anatomy, De humani corporis fabrica in 1543[6]. Because of the Biblical story of Adam and Eve, several initial anatomists contended that "men's ribs" would be lesser in number. The extraordinary hard work of Vessalius carried the flood of the dispute to a stop. Sclerotome cells develop from the costal process of thoracic vertebrae and produce the bony component of each rib. Costal cartilages form from sclerotome cells that migrate into the lateral plate mesoderm via the lateral somatic boundary[7]. Ventral extension and expansions of sclerotomies mesenchyme produce the vertebral arches. The cervical, lumbar, and sacral areas all have these extensions, but the thoracic region does not. The whole primordial costal arch endures chondrification and consequent ossification in the thoracic zone, resulting in the formation of the ribs[8]. In terms of the actual number, of ribs might diverge. It can have less than 24 costal cartilages, bifurcated anterior or posterior ends, synostosis, and diseases such as exostoses or post-traumatic healed callus. Rib aberrations can be congenital or developmental, and they can be the reason for several systemic illnesses, including Gorlin-Goltz syndrome[9].

If there is no abnormality present, the thoracic cage has 12 thoracic vertebrae. These vertebrae are divided into two groups: atypical (first, ninth, tenth, eleventh, and twelfth) and typical (second, third, fourth, fifth, sixth, and eighth). The primary curvature of the thoracic spine, which is ventrally concave and dorsally convex, is known as "kyphosis." The thoracic spine usually retains the primary curvature, which is ventrally concave and dorsally convex, i.e. it is somewhat kyphotic. The twelve thoracic vertebrae begin to grow from intraembryonic mesoderm throughout the third week of intrauterine development [10]. The paraxial, intermediate, and lateral plate mesoderm are discrete layers of the intra-embryonic mesoderm layer. The paraxial mesoderm is systematized as a longitudinal section on either side of the notochord and developing neural tube. The emerging optic capsules split the paraxial mesoderm into preoptic and optic capsules. The post-optic paraxial mesoderm fragments into 40 to 45 segments between the 20th and 30th day of intrauterine development. These segments, known as somites, are prepared craniocaudal[10]. The twelve thoracic vertebrae are made by the sclerotomes, which are ventromedial parts of all 12 thoracic somites. Sclerotome cells distinguish into mesenchymal cells.

The thoracic cage found in a healthy person is made up of 12 thoracic vertebrae, 12 pairs of ribs, and a single-segmented sternum. At costotransverse joints, the tubercle of the rib articulates with the transverse process, and rib heads posteriorly articulate with the conforming vertebral bodies with the aid of the head at the costovertebral joint. Only the first to seventh ribs have vertebrosternal ribs, which have direct costal cartilages that articulate with the sternum anteriorly. The 8th, 9th, and 10th ribs all articulate with the vertebral column posteriorly but do not essentially meet the sternum anteriorly; as an alternative, their costal cartilages fuse and then come in contact with the 7th rib's costal cartilage.

Usually, the thoracic cage has 12 thoracic vertebrae. These vertebrae are divided into two sets: atypical (first, ninth, tenth, eleventh, and twelfth) and normal (second, third, fourth, fifth, sixth, and eighth). The primary curvature of the thoracic spine is the little kyphosis, which is dorsally convex and ventrally concave. The thoracic spine usually preserves its initial curvature, which is ventrally concave and dorsally convex, i.e. it is to a certain degree kyphotic. The twelve thoracic vertebrae instigate to arise from the intraembryonic mesoderm during the third week of intrauterine development[10]. The intraembryonic mesoderm is divided into three sections: (i) the paraxial, (ii) the middle, and (iii) the lateral plate mesoderm. The paraxial mesoderm is organized in a longitudinal column on either side of the notochord and growing neural tube. The growing optic capsules split the paraxial mesoderm into preoptic and optic capsules. The post-ocular paraxial mesoderm split into 40 to 45 segments between the 20th and 30th day of intrauterine development. These segments are identified as somites and are placed in a craniocaudal order[10]. The twelve thoracic vertebrae are molded by the sclerotomes, which are the ventromedial sections of each of the 12 thoracic somites. Sclerotome cells segregate into mesenchymal cells.

During the fourth week of intrauterine life, the sclerotome's free mesenchymal cells glide on all sides of the spinal cord and the notochord to amalgam with cells from opposing somites on the other side of the neural tube[11]. This mesenchyme extends crosswise within the body wall, in such a position that is occupied later by the transverse process, and ventrally within the body wall, a position occupied later by ribs. In the sclerotome section of each somite, the caudal half of each sclerotome advances and joins the cephalic half of each sub-adjacent sclerotome, thereby resulting in re-segmentation. In this fashion, each vertebra is supposed to be built up of the caudal half of one somite and the cranial half of neighboring somite.

All through during early fetal life, the vertebral column and spinal cord raise in an extremely specific and consecutive manner. If any one of the segmentation procedures is disrupted, the vertebral column may suffer from congenital malformations. It is a problematic process to generate and regroup segmental sclerotomes into final vertebrae. Subsequent vertebrae normally fuse asymmetrically or are crowded together. The most common thoracic vertebral body abnormalities are hemivertebrae. The condition in which two vertebral bodies develop normally but do not fuse. The vertebral body is hence made out of the two hemivertebrae. Anterior spina bifida develops when enough space is available between two halves for meninges and nerves to expand forward between them[12].

Vertebral incongruities at one or more levels could give rise to chest abnormalities. In the overall population, hyperkyphosis is the common most thoracic spine disorder. The thoracic spine is typically kyphotic, but kyphosis turns when to hyperkyphosis when it exceeds 400 degrees[13]. It can affect one or more levels of the vertebra and be triggered by several factors[14]. Though juvenile kyphosis happens from birth, it is not apparent clinically till adolescence, between the ages of 14 and 18 years.

Females were found to be more prone to juvenile hyperkyphosis in comparison to males[15]. Kyphosis, like osteochondritis dissecans[16], can be the result of a disruption in the rim epiphysis formation of the vertebral body.

Even though the exact causes of kyphosis are still not known, it is typically presumed that scoliosis is instigated by anterior endplate lesions. Endplate lesions produce intervertebral disc herniation protruding into surrounding vertebral bone, thereby resulting in Schmorl's nodes[17]. The hump prohibited the vertebral ring epiphysis from developing correctly, ensuing about 5 degrees of anterior wedging of the vertebral body. Thoracic hyperkyphosis commonly has its peak at the T7-T9 vertebral level and is known as Scheuermann's disease.

Scoliosis of the thoracic spinal column could pose an effect on the thoracic cage size. Scoliosis is defined as a lateral curvature of the spinal vertebral column that surpasses 10 degrees[18]. Prolonging straight lines across the vertebral body's upper surface when the curve changes direction, infers that the vertebrae that slant the most into the curve's concavity might be utilized to guesstimate the degree of scoliosis. After that perpendicular lines are drawn, Cobb's angle is formed by two perpendicular lines[19]. Anatomical scoliosis is present often from birth and articulates as a flawed structure, such as hemivertebra. If not treated efficiently, anatomical scoliosis can end up in reformed cosmetic anomalies, diminished pulmonary function, and damage to the sensory and motor function of the lower limb, which is known as paraplegia[20]. The second kind of scoliosis was investigated by an unknown etiology, which accounts for 90% of instances. Only a few academics projected that around 4.5 percent of school-age children might get impacted[21].

Scoliosis is classified into three types based on when it appears. Infantile scoliosis is diagnosed between 0-4 years. Group II scoliosis advances between the ages of 4 and 10 years. Adolescent scoliosis is seen to develop after the musculoskeletal development, age, or after ten years. Severe teenage scoliosis rarely emerges in adulthood, and it most regularly affects females 20- 40 years of age. Critical teenage scoliosis is gender-linked since it affects females four times more than males, though less apparent scoliosis with deviation affects both genders at an equal rate[21]. In childhood scoliosis, 90 percent of the bending is towards the left, while in maturity onset scoliosis, the bending is mostly towards the right[22].

Anatomists are fascinated by spinal column irregularities, but health clinicians should be aware of them since they can be the reason for pain, weak muscles of the afflicted limbs, restricted movements, neurological injuries, and transformed sensory perception[9]. "The cervical region is observed as highest out of all regional congenital spinal column abnormalities, succeeded by the thoracic and lumbar areas"[23]. The front sternum bone is located in the midsagittal plane of the thoracic cage and ends bilaterally osteoelastic arches by interrelating with ribs via costochondral cartilages from the first to the seventh ribs. This segmented structure consists of three elements: the cranial prosternum, intermediate mesosternum or gladiolus, and the caudal xiphoid process[24]. Male sternums are characteristically longer in comparison to female sternums[25]. The sternum is normally tilted downward and slightly forward. It is found to be smooth and convex in front, concave at the back, broadest at the first costal cartilage junction, and narrowest at the manubriosternal joint. The location and number of ossification centers differ depending on the degree of completion and timing of sternal plate fusion, along with the breadth of adult bone[26]. The sternum ossification and embryonic development show that it is not linked to the ribs in terms of embryonic genesis[1].

One to three (1-3) ossification centers for manubrium ascend during fetal development during the fifth month of gestation[27]. After that, one ossification center for each of the first and second sternebrae occurs by the fifth month, third and fourth sternebrae appear by the sixth month[26]. The ossification center for the xiphisternum emerges by the third year of life. A sternal foramen or sternal cleft occurs when two ossification centers fail to merge in the midline. The sternal foramen may be related to ectopia cordis, a rare congenital condition[28]. The sternal foramen was first observed in 1649, and it was later recorded in 1707[29]. The frequency of sternal foramen varies from 2.5 to 13.8 percent in the general population[29].

Pectus excavatum (PE) and pectus carinatum (PC) are the utmost common congenital sternal anomalies[30]. PE was found in 23 out of 10,000 infants[31]. Pectus Excavatum (PE) is more commonly found in males than in females[32]. Pectus Excavatum (PE) develops as the sternum is sunk intensely in the thoracic cavity[33], affecting the ribs on either side to protrude more anteriorly than the sternum. The backward sinking of the sternum and cartilaginous segment of the ribs leads to condensed anteroposterior space or diameter of the thoracic cavity. Depending on the degree, PE might appear with minor symptoms and aesthetic issues or with distressing cardiopulmonary problems[34,35]. The heart shifts towards the left and rotates axially as the prevertebral space is condensed. The inward distortion of the sternum might produce right-sided cardiac compression and restrictive pulmonary deficits[36,37]. PE is assumed to be inherited in many factorial ways, while the genes concerned have yet to be recognized[38,39,40]. According to a few researches, PE is uncommon among African Americans and Latinos[41]. Tricherbrust chest is a substitute term for funnel chest. PE has been related to scoliosis and connective tissue diseases such as Ehlers-Danlos syndrome, Marfan syndrome, and Noonan syndrome[42, 43, 44].

Pectus Carinatum is another thoracic malformation known as "pigeon chest" or "keel chest." Its occurrence is similar to Pectus Excavatum, and the sickness affects around one in every 1000 newborns, with men being more afflicted than females[45,46]. The most prevalent concurrent linked illnesses that induce bronchial and pulmonary symptoms are asthma and chronic bronchitis, which impact 16.4 percent of patients[47, 9]. Because adolescence is the most typical time for this illness to be discovered. Parents are usually concerned, and they indicate that the problem arose in the previous few weeks as a result of trauma or weight loss in their kid or teenager[48]. Kyphosis is present in almost all patients, to a greater or lesser degree. Convex everted sternum and costochondral joints are features of Pectus carinatum.

Keel chest/pectus carinatum is commonly seen by the age of ten years, and it gets more pronounced during puberty and adolescence, peaking at the ages of sixteen and eighteen years in females and boys, respectively[49]. The probability of inborn causes increases if pectus carinatum and other thoracic abnormalities run in families[49,50]. Pectus carinatum has been observed to be most common at birth in monozygotic twins[47,48,49]. Marfan's syndrome has been associated with congenital heart disease and delayed hand development[46,51]. Marfan's syndrome has also been associated with valve prolapse, reflux, and mitral stenosis, according to studies[52]. The abnormality is thought to develop as a result of disproportionate excess growth of the ribs' costochondral cartilages[53].

In 1896, Professor Antoine Marfan[54], a French pediatrician, described Marfan's syndrome in a 5-year-old female child. Marfan's syndrome is an uncommon connective tissue genetic illness with an autosomal dominant inheritance pattern and a wide range of severity. Two to three persons out of every 10,000 are affected by Marfan's syndrome[55]. In Marfan's illness, fibrillin, a 350-kD glycoprotein that is the primary structural component of micro-fibril, is abnormal. Marfan syndrome is caused by a genetic defect. It has an "autosomal dominant" inheritance pattern, which means it affects both men and women equally and may be inherited from only one parent with Marfan syndrome. Patients with Marfan syndrome have a tall stature and a thin appearance.

Their fingers, toes, and limbs may look out of proportion to their torso. Their spine may be curved, and their sternum may protrude or be deeply depressed, resulting in less prevertebral space. Their joints may be fragile and easily dislocated. Marfan syndrome is characterized by a long, narrow face and a higher-than-normal roof of the mouth, resulting in crowded teeth[56]. Among the organ systems affected are the bones, eyes, heart, lungs, and blood vessels. Among the organs injured are the bones, eyes, heart, lungs, and blood vessels. There are several excellent explanations of typical features in both adults and children[57] can be found. The Ghent nosology[58] major and minor criteria are used to diagnose Marfan's syndrome.

The prevalence of Ehlers-Danlos syndrome in the general population, including all variants/subtypes, is best estimated to be between 1 in 2500 and 1 in 5000[59]. Ehlers-Danlos syndrome (EDS) is an intriguing group of hereditary collagen and extracellular matrix protein abnormalities that present in various ways. Although Ehlers-Danlos Syndrome is most typically inherited as an autosomal dominant trait, it can also occur as a de novo mutation in up to 50% of people[60]. The most important clinical issues include withering scar development, very stretchable skin, and extensive joint hypermobility. The fold of the top eyelid covers the inner corner of the eye, the skin is easily breakable, mushy, and "doughy," there is easy bruising, herniation, joint hypermobility difficulties, molluscoid pseudo-tumors, subcutaneous spheroids, and a first-degree relative is afflicted. Minor clinical criteria for Ehler-Danlos syndrome diagnosis and classification are used more frequently[61].

The disease's physical manifestations include (i) extremely stretchable skin, smooth and velvety texture, britality of skin, delayed wound healing, and thin atrophic scars after wound healing[60], (ii) hypermobility of the joints, resulting in repeated subluxation and dislocation, which can lead to early osteoarthritis and chronic pain[62], and (iii) classic facies, which include down slanting palpebral fissures, epicanthal[61] 3 iv) Cardiovascular symptoms include mitral valve prolapse and, less frequently, tricuspid valve prolapse, and vi) Hypotonia, which may appear as delayed motor milestones such as walking in the developing infant, is one of the neurological symptoms. Generalized discomfort, sleeplessness, and persistent exhaustion are also frequent[61]. Minor bone anomalies, such as pectus excavatum, are prevalent, particularly in individuals with type IV illness[63].

The prevalence and frequency of Noonan Syndrome at live birth are estimated to be one in 1000-2500 persons globally[64,65]. As a result, trisomy 21 chromosome is the second most prevalent cause of congenital heart illness[64,65,66]. Though autosomal dominant inheritance accounts for the vast majority of Noonan syndrome patients, autosomal recessive inheritance has been documented in a few instances[67]. Noonan syndrome is characterized by slowed growth after birth, distinctive facial dysmorphism, a wide range of congenital heart defects (CHDs), learning difficulties, short height, renal anomalies, lymphatic malformations, bleeding disorders, and skeletal deformities [64,68,69]. Two major clinical features of Noonan Syndrome include atrioventricular canal partial defect[70] and mitral valve insufficiency in the cardiovascular system. Hyperelastic skin, multiple lentigines, nevi, thick curly hair or thin sparse hair, low posterior hairline with webbed neck, thick curly hair or thin sparse hair, low posterior hairline with webbed neck in integumentary system and its appendages[71]. The supraclavicular area of the facioskeletal system revealed articulation trouble, a high arched palate, malocclusion, micrognathia[72], and a hearing abnormality[73]. Scoliosis, pectus excavatum, and pectus carinatum are all skeletal system anomalies[74].

Since the inception of the thoracic cavity as an entity, beginning with the end of the embryonic stage (the 8th week of intrauterine life of development) and continuing until birth and adulthood, how the thoracic cage assumes its shape and how its diameter grows and interacts with each other is unknown. What is the pattern of growth, differentiation, and development that occurs during the prenatal, postnatal, adolescent, or growth spurt, adult, and illness condition? It is the subject of discussion and investigation. A thorough examination of textbooks reveals two types of human development patterns during gestation and postnatal life.

The two forms of growth patterns in the human body are isometric growth patterns and allometric development patterns. Differences in the relative rates of growth of one bodily component vs another are referred to as allometric growth. Allometric growth is observable in fetuses, neonates, toddlers, and adults. True isometric growth would indicate a proportionate rise in all organs and systems throughout time. Isometric growth does not appear to occur in developing embryos[75]. Humans develop proportionally in all areas of the body after birth. Body length rises from 48-53cm at birth to roughly 75cm in the first year, and by 12 to 13cm in the second. Each year, an additional 5 to 6cm is added. Individual longitudinal development curves show a rise in development velocity from 10.5 to 11 years for girls and 12.5 to 13 years for boys.

This rapid rise in size is known as an adolescent growth spurt. This spurt in growth happens every 2 to 2.5 years in both sexes. Females acquire around 16 cm in height during the growth spurt, with a peak velocity at 12 years of age. Boys grow about 20 cm in height, owing mostly to trunk growth. Boys attain a peak growth rate of 10 cm per year when they are 14 years old. Height growth decreases after the teenage growth spurt, and observable growth is thought to halt at roughly 18 years in females and 20 years in males. A longitudinal study shows that the average age for girls is 16.25 years and 17.75 years for boys, with a usual deviation of +/- 2 years[76]. Almost every skeletal and muscular structure is affected by the spurt, although not in the same way. The majority of the height growth is due to an increase in trunk length rather than an increase in leg length. The dimensions increase in a very regular fashion. Leg length is generally the first to peak, followed by torso breadth, and finally the shoulder.

The thoracic cage's anteroposterior, transverse, and vertical dimensions grow in proportion to a child's age after birth. It is still unknown how the thoracic cage develops in three dimensions. It is unknown what the normal connection between transverse diameter and anteroposterior diameter is at the most proximal and distal regions of the thoracic level[77]. How does the connection between transverse and anteroposterior diameter change with age in North Indians, and is there a gender difference? Furthermore, it is unknown how the connection between these two diameters varies with developmental spurts and how it behaves in adulthood in North Indians.

In 1987[78], Haller and his colleagues established a technique for quantifying the connection between transverse and anteroposterior dimensions in pectus excavatum patients to classify them as surgically correctable or non-interventional. The Haller index was calculated by subtracting the transverse and minimum anteroposterior internal thoracic diameters from an axial Computed Tomography thorax picture. This index was created to assist clinicians in making treatment decisions for patients with pectus excavatum. Measurements were collected from an axial Computed Tomography thorax slice at the level of maximum deformity. The distance between the posterior surface of the sternum bone and the anterior surface of the body of the thoracic vertebra at the same vertebral level was used to determine the anteroposterior diameter. Even though the sternum bone protrudes laterally concerning the vertebral column, these two internal transverse and anteroposterior dimensions are measured such that they intersect at a right angle. When calculating landmarks, any soft tissue above the bones is never taken into account.

Daunt and his colleagues assessed the Haller index in 2004[79] to define thoracic measures for people who do not have a pectus excavatum abnormality of the thorax. This time, each person was assessed at a distinct spinal level, and each vertebral level was unique. In 2016, James E. Archer and colleagues utilized Haller's methodology to characterize the size of the thoracic cavity in the British population from T3 to T12. Is it applicable to other groups, such as north Indians, because it is rarely explored and studied in Indians? Height, weight, BMI, gender, and ethnicity all have their unique genetic makeup and physical body build in Indians as compared to other geographically distributed races. As a result, a study into the anticipated pattern of growth ratio per-vertebral thoracic level concerning age and gender in diverse geographically scattered human populations is critical.

Clinicians frequently observe that long-standing diseases of the lung, heart, and space-occupying thoracic lesions can modify the thoracic dimensions. According to reference texts of medicine such as Harrison and Davidson, obstructive pulmonary disease at its ultimate stage or depending on severity deforms the chest. Chronic Obstructive Lung Disease (COPD) is the most common obstructive pulmonary disease, causing pigeon chest or barrel shape chest with variable degrees of breathing difficulty. It is unclear how Haller Index defines these changes. The majority of the studies using Haller's approach at various vertebral levels in COPD patients were conducted in other countries. There is relatively little research available in Indians. This technique has such a broad reach that this basic indicator may be used as a tool to anticipate illness development, severity, and prognosis about changing thoracic dimensions.

Coughing and shortness of breath are symptoms of a variety of respiratory disorders, which may be classified into three types: 1-obstructive lung disease, 2-restrictive diseases, and 3-vasculature anomalies. Obstructive lung diseases include asthma, chronic obstructive pulmonary disease, bronchiectasis, and bronchiolitis[80]. Chronic obstructive pulmonary disease (COPD) is an inflammatory lung disease. Chronic bronchitis and emphysema are the initial and terminal phases of chronic obstructive pulmonary disease (COPD) respectively, and they frequently overlap. It is defined as a chronic illness characterized by persistent respiratory discomfort and restricted airflow that is not completely reversible. Emphysema is a chronic obstructive pulmonary disease marked by the loss of lung alveoli and the expansion of the air space. Chronic bronchitis is a medical disorder distinguished by a chronic cough, phlegm, and involvement of small airway narrowing and thickening. Chronic bronchitis is distinguished by a reduction in the number of small bronchioles as well as their constriction[81].

Patients with COPD in their early stages usually have a fully normal physical examination. Physical examination of the lungs in people with more severe illnesses indicates a longer expiratory phase, which may include wheezing. Percussion can also be used to identify hyperinflation symptoms such as a barrel chest and increased lung volumes with the poor diaphragmatic excursion. By sitting in the traditional "tripod" position, patients with significant airflow obstruction may also engage supplemental respiration muscles such as the sternocleidomastoid, scalene (scalenus anterior, scalenus medius, and scalenus posterior), and intercostal muscles[82].

According to conventional knowledge, patients with severe emphysema termed "pink puffers," are slim and lack blue pigmentation over the nail bed, lips, and tongue at rest, with significant use of auxiliary muscles. Patients with chronic bronchitis, on the other hand, are more likely to develop cyanosis and are referred to as "blue bloaters." [82]. Advanced disease may be accompanied by cachexia, substantial weight loss, bitemporal wasting, and widespread subcutaneous adipose tissue loss. This sickness has been associated with both inadequate oral intake and increased levels of inflammatory cytokines. The 'Hoovers sign'[83,84] is a paradoxical inward displacement of the rib cage with inspiration that happens in certain very unwell people. The vector of diaphragmatic contraction on the rib cage has been changed as a result of hyperinflation. Signs of overt right heart failure, termed as 'cor-pulmonale,' have been significantly less prevalent since the development of supplemental oxygen treatment.

If a patient coughs up phlegm on most days for at least three months in a row for more than two years, bronchiectasis and untreated bronchial asthma are ruled out[85], and the patient is diagnosed with chronic bronchitis. Chronic bronchitis arises from recurrent contact with irritants such as cigarette smoke, tobacco smoke, dust, smoke, and fumes on the bronchial mucosa. Exacerbations of chronic bronchitis can be exacerbated by moisture, fast temperature fluctuations, and fog. Chronic bronchitis is a common illness in middle and late adulthood, and smokers are significantly more likely to get it than nonsmokers. People who live in cities are more prone than those who live in rural areas to get chronic bronchitis. The diseases that induce sickness include hypertrophy of the mucous-secreting glands and an increase in the number of goblet cells in the bronchi and bronchioles, with a consequent reduction in ciliated cells. Mucosal edema and prolonged structural damage to the airway walls diminish the diameter of the airways. A considerable portion of the airflow obstruction in chronic bronchitis is permanent.

Because the degree of constriction is greater during expiration, the air is trapped in the alveoli. Excessive alveolar distension causes alveolar wall disruption. Chest X-rays are typically uninteresting; however, they might reveal symptoms of emphysema. Chronic bronchitis consequences include respiratory failure, pulmonary hypertension, and secondary polycythemia. It is a chronic illness with acute exacerbations and remissions that leads to right-sided heart failure and respiratory failure. A chest deformity known as "pigeon chest"[85] occurs in the most severe instances.

Emphysema signifies inflation in the concept of abnormal distension of alveoli with air. It is a medical disorder characterized by the aberrant presence of air in body tissue on occasion. Smokers are more prone to get emphysema. By producing persistent inflammation in the airways and alveolar septa, smoking is hypothesized to cause irreparable damage to the supporting connective tissue and alveolar septa, particularly at their attachments to the small airways. Alpha-1 antiproteinase deficiency is a rare genetic disorder that produces severe emphysema in adolescents. Alveolar damage may be classified into two types. There are two categories of injuries based on the severity of the injury. (I)-panacinar emphysema, characterized by signs of widespread alveolar wall degeneration, and (II)-centrilobular or centriacimar emphysema, which mostly affects alveoli that are tightly related to the respiratory bronchioles.

Clinical features of advanced emphysema include (i) a reduction in tracheal length palpable above the suprasternal notch, (ii) downward movement of the trachea during inspiration known as "tracheal tug," (iii) scooping of supraclavicular and suprasternal fossae during inspiration, (iv) jugular venous filling in inspiratory phase, and (v) withdrawal of coastal margins and intervene[86]. When chest radio-imaging done, the images/skiagrams, especially X-ray chest delineates emphysematous bullae, are evaluated, hyper translucent lung fields with loss of peripheral vascular lines, terracing of the hemi diaphragms, and massive pulmonary shadows at both hila are seen. Although high-resolution computed tomography (HRCT) may reliably identify emphysema, it is rarely used for routine diagnostic reasons[86].

Bronchial asthma is a kind of obstructive pulmonary illness. Asthma symptoms are caused by inflammatory responses that occur inside the bronchial wall and include numerous cells, mediators, and cytokines, as well as eosinophils, mast cells, lymphocytes, and other cells. An inflammatory condition that causes the bronchi to become hypersensitive to various stimuli, leading them to constrict rapidly. Asthma is characterized by inflammatory cells, plasma exudation, edema, smooth muscle hypertrophy, mucous obstruction, and epithelial shedding in the airway. Asthma symptoms include coughing, wheezing, chest tightness, and dyspnea, which are often worst at night. There are two forms of asthma: (a) early-onset asthma, often known as atopic[87]. The first type of asthma, which often begins in childhood and affects atopic persons who rapidly produce IgE antibodies to regularly encountered allergens, and (b) late-onset asthma[87], which can occur at any age and affects non-atopic individuals. Bronchial asthma can be intermittent, severe acute, or status asthmaticus, a condition marked by life-threatening asthma episodes as well as chronic asthma.

Severe asthma is characterized by a pulse rate of more than 120, pulsus paradoxus (difference between heart and pulse rate), and difficulty in speaking sentences. Life-threatening symptoms include the inability to speak, central cyanosis, fatigue, confusion, decreased awareness level, quiet chest, and unrecordable peak flow. Radiological imaging displays a normal appearance in between episodes; nevertheless, in long-term instances, the appearance may be indistinguishable from emphysema-induced hyperinflation[87]. Pigeon chest deformity has been observed in long-term instances[87]. When tenacious mucus blocks a large bronchus then there is opacity owing to lobar or segmental collapse, high resolution computed tomography (HRCT) thorax can be used to diagnose the problem properly.

Bronchiectasis is defined as an abnormal permanent dilation of the bronchi[88,89]. It can be caused by several factors. Bronchiectasis is most usually induced in children by a severe bacterial disease such as whooping cough or measles. Bronchiectasis is classified into three categories based on the cause. Congenital hypogammaglobinemia, cystic fibrosis, and primary hypogammaglobinemia; pneumonia, whooping cough, primary tuberculosis, and foreign bodies in children; and suppurative pneumonia, pulmonary tuberculosis, pulmonary eosinophilia, and bronchial tumors in adults. In adults with primary ciliary dyskinesia, Kartagener syndrome and Young's syndrome are uncommon[89]. Bronchial distension is caused by pus accumulation beyond a lesion blocking a major bronchus in this situation.

Granulation tissue, squamous epithelium, and typical Ciliated epithelium line the bronchiectactic cavities. Symptoms include (i) chronic cough and copious phlegm production, which is usually worse in the morning and is often caused by posture changes; (ii) fever, malaise, and an increase in coughing and squamous sputum production; (iii) recurrence of pleurisy, which often occurs at the same site; and (iv) slight or massive blood, which is usually recurrent. If the condition is severe, the patient will have constant feculent and copious sputum, weight loss, anorexia, weariness, sleep sweating, and inability to develop in children, all of which will cause a general decline in health. Bronchiectasis is typically not seen on standard chest x-rays unless it is severe. Bronchography or high-resolution computed tomography (HRCT) is virtually generally used to detect bronchiectasis at the moment[89]. High-Resolution Computed Tomography's main observable diagnostic findings are: (1) an internal bronchial width greater than the neighboring pulmonary artery; (2) a bronchial failure to taper; and (3) an exterior bronchi visualization of 1-2cm of the lung fields[90].

Chronic obstructive pulmonary disease (COPD) is a catch-all name for a group of illnesses with comparable causes and progression, signs and symptoms, and morphological consequences. It is among the top cause of death in the globe[91]. This apparent variability in the disease's appearance, characteristics, the severity of impairment, and future prognosis impedes the correct identification of smaller groups or juvenile categories of populations that are at risk for a quick breakthrough, defeating treatment planning. To present, the bulk of research on chronic obstructive pulmonary disease (COPD) has involved persons with an average age of more than 60 years[91]. The most prevalent signs of the chronic obstructive pulmonary disease include airflow limitation in the respiratory bronchioles, which is accompanied by a long-term cough with sputum expectoration, breathlessness in exercise, and a distinctive lung sound known as wheeze[92]. By 2030, Chronic Obstructive Pulmonary Disease (COPD) is expected to be the world's third greatest cause of death[93].

Chronic obstructive pulmonary disease is a lung disease caused by a complicated biochemical and cellular sequence in the bronchioles and neighboring alveoli[94]. The most noticeable features of chronic obstructive pulmonary disease are alveolar damage and alveolar hyperinflation caused by trapped air (emphysema). Histological slices of emphysematous lung tissues evaluated under a scanning electron microscope for alterations in the histologic structure show significant lowering and disintegration in the normal form and arrangement of alveoli, as well as inflated alveolar ducts. The vast majority of people with the chronic obstructive pulmonary disease have decreased elastic recoil, which, together with decreased expiratory air flow, leads to alveolar overinflation and disease progression[95]. Overinflation of the lungs in individuals with chronic pulmonary lung disease alters the geometry of the chest wall and the structure of the thoracic cage, leading to thoracic wall dysfunction. Some studies conducted in Western nations on individuals with the chronic obstructive pulmonary disease have demonstrated changes in form and size, as well as internal thoracic measurements[96,97]. Only a few studies in India have looked at the impact of COPD on thoracic dimensions. Exploration of thoracic cage deformity is therefore clinically significant because, in this contemporary era of medical technology, there is a suggestion of lung transplantation in severe situations.

In the middle of chronic obstructive pulmonary disease, the diaphragm is comparably flatter and at a lower level due to excessive inflating of the lungs as a whole, and the magnitude of juxtaposition is reduced compared to normal persons. As a result, the distal rib cage expansion excursion induced by diaphragmatic contraction in COPD patients is frequently smaller than in healthy persons. Normally, the muscle fibers of the diaphragm are oriented cranially at their origin on the ribs; however, in COPD, the muscular fibers of the diaphragm are directed transversely inwards. In this circumstance, diaphragm recession causes a reduction in the transverse width of the distal thoracic cage or the volume filled by the constricted lower rib cage[98,99,100]. Many architectural manifestations inside the lungs, such as emphysema, airway remodeling, mucoid secretion obstruction, and vascular remodeling[101,102,103], are likely to result in the clinical diversity of the chronic obstructive pulmonary disease.

As indicated by the preceding account, the thoracic cage dimensions play an important role in the development of chronic obstructive pulmonary disease and how to regulate breathlessness in COPD patients. Based on changes in thoracic dimensions in healthy persons and COPD patients, improved chronic obstructive pulmonary disease management may be achievable in the absence of pulmonary function testing, allowing COPD patients to have a higher quality of life. As a result, the current study's major purpose is to compare thoracic cage measures in healthy persons and COPD patients.

Patients in medical practice regularly visit doctors for cardiorespiratory symptoms or thoracic cage difficulties. An X-ray chest is the most commonly recommended diagnostic method for tracing the normal and diseased architecture of the thoracic cage. An X-ray chest is still the gold standard for diagnosing a variety of thoracic disorders and anomalies because it is inexpensive and widely available. If doctors are unclear, a thoracic computed tomography scan may be utilized to confirm the diagnosis and answer questions. Computed tomography of the thorax provides physicians and surgeons with cross-sectional anatomy and greater detail. Radiologists, orthopedists, cardiothoracic surgeons, pediatricians, pediatric surgeons, physicians, and cosmetic surgeons all need to understand the thoracic cage architecture.

The thoracic cage houses the right and left lungs, as well as the heart. To function appropriately, optimally, and efficiently, these critical organs require proportional internal space in the thoracic cavity. Internal thoracic cavity space and volume are largely governed by internal transverse and anteroposterior diameters, as well as vertical diameter, which must be acceptable in normal healthy persons. This also demonstrates a person's ability to work healthily and productively. Changes in the diameters of the thorax are caused by congenital or acquired abnormalities of the ribs, thoracic vertebrae, and sternum. Aside from these skeletal element deformities of the thoracic cage, some long-standing lung, heart, and space-occupying lesions are known to impose their effects and alter the geometry of the thorax in humans. Emphysema, chronic bronchitis, long-standing bronchial asthma, and bronchiectasis are the most prevalent lung diseases that induce pigeon chest deformity.

Nonetheless, because the anomalous thoracic cage was assessed at one unnamed thoracic vertebral level, many unique characteristics of the abnormal thoracic cage are not addressed using Haller's methods. Initially, Haller evaluated the transverse (width) and anteroposterior (height) dimensions of the thoracic cage at the level of greatest deformity in pediatric patients, presuming that sternum abnormality was a major contributor to pectus excavatum or carinatum. Because various patients had varying vertebral levels, it was unclear which section of the sternum was most usually implicated in causing the aforementioned deformity. Which thoracic segment or vertebrae are most commonly implicated in thoracic cage abnormalities? Are rib anomalies accountable for changes in the morphology and internal diameters of the thorax? How age, gender, and ethnicity influence the development of a normal individual without deformity. Can long-term lung problems, such as chronic obstructive pulmonary disease (COPD) or thoracic space-occupying lesions, impact the shape?

The current study is done to examine axial Computed Tomography thorax pictures of normal persons and chronic obstructive pulmonary disease (COPD) patients from North Indian population samples for structural alterations in the thoracic cage to discover answers to the above issues. By applying Haller's methodology in both normal subjects and patients of COPD the Haller's index was determined at each vertebral, Starting from thoracic vertebral level T3 to twelfth thoracic vertebra in craniocaudal direction by measuring maximum internal transverse diameter and minimum ateropterior diameter radio-graphically.

WHY THIS STUDY HAS BEEN CONDUCTED?

1. Initially Haller index was measured in paediatric patients for descriptive and decision-making tool for the management of pectus excavatum in 1987. It was a case report. Not in adults. Present study is intended to measure in adults.
2. Haller et al. (1987) and subsequently Daunt et al. (2004) have used the index to describe thoracic dimensions in those without a pectus deformity, but this has only been at one vertebral level per subject, which was not always the same vertebral level in all patient. Again, this study included paediatric population in abroad. Present study is intended to measure and determine at fixed multiple vertebral levels.
3. J. E. Archer et al (2016) first time studied British paeditric population between 8 to 18 years without chest deformity at 10 fixed vertebral levels from T3 to T12. Again, study participants were not adult.
4. Haller index at multiple vertebral levels (T3 to T12) rarely been studied in adult population in Indians or in abroad and if studied it was at few selected vertebral levels (2 to 3). In caudal thorax at xiphisternal joint level was studied only but not below this.
5. In COPD patients Haller index was studied keeping spirometric value for severity of disease and not the anatomically identifiable change in lung parenchyma on HRCT findings. Moreover in previous studies the vertebral levels were limited to two or three
6. In COPD multiple vertebral levels were rarely studied.
7. Study of Haller index in Indians in early adulthood (18) to old aged population to determine the effect of aging rarely been studied in nondeformed chest.

RESEARCH QUESTION

Can the Haller index be used as a tool to decide the chest deformity at different vertebral levels in normal adult people and chronic obstructive pulmonary disease patients?

RESEARCH HYPOTHESIS

1. Long standing diseases of the lungs could affect transverse and anteroposterior diameters of the thoracic cavity in chronic obstructive patients.
2. Alternate null hypothesis "Long standing diseases of the lungs could not affect transverse and anteroposterior diameters of the thoracic cavity in chronic obstructive patients".

AIM

To measure and compare the thoracic dimensions at multiple vertebral levels in normal and chronic obstructive pulmonary disease patients using computed tomography thorax images in axial plane.

OBJECTIVES

1. To determine dimensions of thoracic cage radiographically in the Indian adult population by measuring internal maximum transverse and minimum anteroposterior diameters of the normal and Chronic Obstructive Pulmonary Disease patients, using axial computed tomography-thorax images at measurable thoracic vertebral levels.
2. To calculate the ratios of transverse and anteroposterior internal diameters in the cranio-caudal direction from vertebral level T3 to T12 for CT-thorax images of Indians.
3. To determine the Haller Index in normal and Chronic Obstructive Pulmonary Disease patients in Indians.
4. To find an insight that could help early detection and better management of Chronic obstructive pulmonary disease patients.

REVIEW OF LITERATURE

Surendran S et al., 2022[104] used a geographic information system (GIS) in the semi-urban area of Trivandrum, South India, to map COPD cases and their risk factors, and to identify associations between them. We conducted a study aimed at determining This community-based, cross-sectional, descriptive study (n = 494) was conducted in a subcenter area of a primary care center. COPD prevalence (6.5%) was close to national estimates. Spatial maps showed that COPD cases were concentrated in areas with high firewood exposure, high smoke exposure, and homes close to local highways. High-risk clusters were identified that had significant correlations with all risk variables. GIS technology can be used to identify geographic clusters and environmental risk factors for COPD patients, making it a valuable tool for the targeted treatment of COPD.

Feizi H et al. In 2022[105] the regional age-standardized point prevalence and mortality rates for COPD in 2019 were 2333.9 (2230.1, 2443.6) and 26.1 (22.2, 29.5) per 100,000, respectively, with 30.6% (28.2%, 33.0%). Since 1990, there has been an increase and a decrease of 18.0% (2.8%, 30.9%). The regional age-standardized DALY rate in 2019 was 649.1 (5746,717.7 per 100,000), a decrease of 11.8% (0.9%, 21.1%) since 1990. Additionally, by gender, smoking (43.7%), particulate matter exposure (22.8%), and occupational particulate matter exposure (11.4%) had the highest proportion of DALYs attributed in 2019. COPD is the leading cause of mortality and disability in the MENA region. Nevertheless, COPD is responsible for a significant number of deaths and DALYs, especially among the elderly. Programs targeting risk factors such as smoking should be considered.

Chen, Y et al. In 2022[106] two-dimensional shear wave elastography (2D-SWE) was used in this study to quantify diaphragm stiffness (DS) and intercostal muscle stiffness (IMS) in patients with COPD. The efficacy of 2D-SWE in measuring respiratory function was also determined. A total of 219 of COPD patients and 20 healthy individuals participated in the study. DS and IMS, and pulmonary function were assessed using 2D-SWE. DS and forced expiratory volume in 1 second (FEV1)/forced vital capacity (FVC), predicted FEV1% value, residual capacity (RV), total lung capacity (TLC), RV/TLC, functional residual capacity (FRC) Inspiratory capacity (IC) was greater in COPD patients (r1 = -0.81, r2 = -0.63, r3 = 0.65, r4 = 0.54, r5 = 0.60, r6 = 0.72, and r7 = -0.41; P < 0.001).) was stronger than IMS (r1=-0.76, r2=-0.57, r3=0.57, r4=0) 47, r5 = 0.48, r6 = 0.60, r7 = –0.33. P < 0.001). 2D-SWE may be used for DS and IMS evaluation. A specific correlation was observed between respiratory muscle stiffness and lung function. DS and IMS gradually increased as the severity of COPD worsened and pulmonary dysfunction progressed.

Chavan SD et al., 2021[107], this study was conducted to compare chest measurements in normal and COPD patients. Two groups of 70 COPD patients and 70 healthy participants were formed. Group I for COPD patients and group II for healthy controls, HDI, pulmonary function tests, and mean transverse rib diameter were also compared. The mean AP diameters in group I were 10.9 mm and 9.1 mm in group II, the transverse diameters in group I were 22.7 mm and 21.0 mm in group II, and the HDI was 23.7 mm in group I and 20.1 mm in group II. FEV1 was 54.1% in group I and 89.2% in group II, and FVC was 47.4 in group I and 91 in the group.

Viswambhar V, Bandlamudi S, Aruna S, Meenakshi NS and others. al. (2020)[108] conducted a cross-sectional study in healthy participants and (COPD) patients in the respiratory department of a tertiary care hospital to monitor changes in chest circumference. Their main concern was to compare the chest cavity sizes of healthy people and COPD patients. Through their studies, they found that patients with no access to pulmonary function testing facilities were aware that changes in chest size were more susceptible to disease progression, and thus improved management of dyspnea could improve the patient's condition. They investigated what could be improved. The study enrolled 80 cases and 80 healthy controls. The results summarized that the mean AP diameter in COPD cases was statistically significant [p<0.001], being [10.64 cm ± 2.16cm] larger compared to healthy controls [9.29cm ± 1.47cm]. The mean transverse diameter of COPD patients was 22.5 cm ± 2.1 cm and healthy controls [21.83 cm ± 2.02 cm statistical significance (P = 0.041). They also observed a mean HDI of 23.35 cm ± 2.6 cm in COPD cases and 20.57 cm ± 0.91 cm in healthy controls, with statistical significance [p<0.001]. As a result, it was concluded that the transverse chest diameter,

AP diameter, and admeasured chest radiography were higher in COPD patients than in healthy controls.

Silva TK, Zanon M, Altmayer S, Pacini GS, Watte G, Stein R, et al. (2020)[109] compared quantitative CT parameters in children with severe asthma and healthy subjects and also their compared clinical characteristics. They performed a retrospective study analyzing correlations between healthy and asthmatic patients. They analyzed their CT data of 19 school-age children (5–17 years) with severe asthma and 19 age-matched control children with PE. CT parameters such as MLD, TLV, AWT, AT%, and AWT% percentages were observed. The results thus obtained showed that the asthma patients had higher mean AT%. The mean AT% for previously hospitalized patients was 29.0 ± 4.7% and for those without previous hospitalization was 19.2 ± 5.0% ($p<0.001$). AT% showed a highly negative correlation with FVC and a moderate correlation of 25-75% with FEF. It was concluded that children with severe asthma had differences in quantitative chest CT scans compared with healthy controls, showing strong correlations with previous hospitalizations and pulmonary function tests.

Mason Uvodich, Ross Barmann, Adam Reitz, et al (2020)[110] In this case report, investigators had studied Marfan syndrome who underwent spinal instrumentation and T3 to L4 fusion to correct symptomatic scoliosis. The patient was described in detail. The patient's blood pressure was controlled by vasoactive drugs and high fluid intake during the complex perioperative course. Persistent hemodynamic instability required intensive care despite postoperative resuscitation. Common reasons for postoperative hypotension were diagnosed as B. Sepsis, hypovolemic shock, prolonged bleeding, and prolonged effects of anesthesia. Finally, the patient's refractory hypotonia was achieved both by sustained intraoperative prone to PE-aggravated mechanical compression and by surgically modified spines that reduced the diameter of the thoracic cavity. Exacerbation of chest wall deformities as a result of mechanical compression of thoracic contents can be an obstacle to the correction of spinal deformities.

Beop-Yong Lim, Youngwoong Kim, Hoseok I et al. (2020)[111] This review describes different types of surgical procedures as part of Nuss surgery, and this study is computational in- vivo. Their aim was to verify their relative superiority by a mechanical method. However, no studies have focused on the impact of the sternoclavicular joint on the Nuss procedure simulation. Therefore, in this study, they aimed to validate the motion of these joints by combining clinical and finite element analysis data. Scenarios were established by classifying joint movements based on translational and rotational constraints in the coordinate plane. The analysis was performed by applying the set scenarios to the constructed finite element model of the parapet. Haller Index Sternal Displacement and Equivalent Stress were compared with postoperative patient data. The results were similar to the actual surgery due to the anterior restriction on chest wall movement. Considering the impact of the sternoclavicular joint was recommended for more accurate results.

Mayke J. van Dort, Johanna HM Driessen, Elisabeth APM Romme, et al. (2019)[112] Investigating the association between ventricular fibrillation and kyphosis severity, and ventricular fibrillation risk in smokers with and without COPD. They conducted a study to of 1239 subjects, 253 (20.4%) had predominant AF and 294 (23.7%) developed AF within 3 years. The number, presence, and severity of prevalent ventricular fibrillation correlated with a greater kyphosis angle. The mean increase in kyphosis angle over 3 years was smaller in subjects with ventricular fibrillation but significantly greater than those without ($p < 0.001$). After adjusting for bone weakening (BA) and predominant ventricular fibrillation, baseline kyphosis angle was associated with the onset of ventricular fibrillation within 1–3 years (+T4 to T12 angle per SD, hazard ratio [HR] = 1.34 [1.12-1 .61] and HR 1.29 [1.15–1.45]).

Malay Sarkar et al. (2019)[113] This review focuses on some physical manifestations, etiology, and clinical significance of chronic obstructive pulmonary disease. Global footprint; patient history taking and laboratory tests are declining day by day. One reason for this may be the tremendous advances in the field of medical technology that has led to an over-reliance on laboratory procedures, advanced diagnostic equipments, and medical tests as the primary modalities of patient management. In resource-poor countries, a detailed medical history and physical examination should be emphasized as an important step in patient diagnosis and treatment. Currently, the decline in physician bedside and clinical skills is certainly a problem. Physically diagnosing

COPD is the quickest step leading to an early diagnosis and treatment. Identifying physical signs at the bedside should be the first step in any diagnostic and therapeutic approach.

Ravi S Reddy, Khalid A Alahmari et al. (2019)[114] conducted a study of 75 subjects, including 25 healthy subjects (nonsmokers, smokers, COPD) aged between 18 - 70 year. Researchers did measurements (2 levels) above and below the chest extension (CE) which were obtained using cloth inch tape. The intra-rater (between-day) and inter-rater (intra-day) consistency of CE measurements was assessed by two assessors. Vital capacity (VC), pulmonary function parameters, forced vital capacity (FVC), forced expiratory volume in one second (FEV1), and FEV1/FVC were measured using a computerized spirometer (Spiro lab 3). It was observed that intra-rater reliability for top and bottom CEs showed a highly valuable association with intraclass correlation (ICC) values ranging from 0.90 to 0.93 for top CEs and 0.85 to 0.86 for bottom CEs. Both upper and lower CE limits were significantly and positively correlated with all pulmonary function parameters and strongly correlated with FEV1/FVC ($r = 0.68$). Else, upper and lower CE measurements taken via. inch tape exhibited good consistency and reproducibility with respect to intra- and interrater in healthy nonsmokers, smokers, and COPD subjects.

Kimia Khalatbari Kani, Hyojeong Mulcahy, and Jack A. (2019)[115] reviewed the anatomy, injury mechanisms, classification, imaging, treatment, and long-term consequences of thoracic osteochondral injury in this review. It was concluded that rib fractures were the predominant type of thoracic impact injury. Multiple rib fractures are a significant sign of trauma severity, and additional rib fractures mainly lead to increased morbidity and mortality in the elderly. Chest injuries are often accompanied by fatal injuries regularly. Accurate recording of the extent, pattern, and severity of acute thoracic injury aids in more accurate imaging of associated injury.

David Lo, Amit Maniyar, Sumit Gupta, and Erol Gaillard (2019)[116] evaluated CT findings in a local cohort of children with severe asthma and evaluated clinical or pathobiological parameters for which CT was predicted to change. This was a retrospective, single-center, observational study of all children attending the Leicester difficult asthma clinic (DAC) who underwent chest CT between 2006 and 2011. In addition, eight age-matched non-asthmatics were included in this study to correlate differences in their CT findings between asthmatic and non-asthmatic children. Two radiologists were recruited to evaluate the CT images in an independent review. DAC patients were further divided into two groups per noted abnormality for comparison with pre-recorded clinical parameters, including sputum white blood cell count, age, total serum IgE level, and lung function. Thirty DAC patients (median 12 years, range 5–16 years) were involved. Bronchial wall thickening (BWT) and air entrapment (AT) are commonly seen abnormalities, observed in 80% and 60% of his DAC patients, respectively. Bronchiectasis (BE) was observed in 27% of DAC patients. DAC patients with BE on CT images were found to be older than DAC patients without BE (13.9 ± 0.67 vs 11.5 ± 0.61, p = 0.038).

SJ Lim et al. (2018)[117] enrolled in their study 85 COPD patients, 76 men, and 9 women. (Mean age 70.6 ± 7.1 years) and 30 normal controls. Chest dimensions were measured at levels 3, 6, and 9 of the thoracic vertebrae. The mean sagittal anteroposterior (AP) diameter, maximum transverse diameter, and maximum AP diameter of the left and right hemithorax were also assessed. The mean AP diameter was observed to be significantly larger in his COPD patients compared to normal controls (13.1 ± 2.8 cm vs. 12.2 ± 1.13 cm; p = 0.001). The AP diameter/thoracic transverse diameter ratio was also significantly higher in COPD subjects than in normal controls (0.66±0.00). 061 vs 0.61 ± 0.86; p=0.002). Also, in patients with COPD, thoracic AP diameter was positively correlated with 6-min test walking distance (r=0.395, p<0.001 and r=0.238, p=0.028) and BMI, and negatively correlated with progression. Age (r=-0.231, p=0.034). Next, multiple regression analysis also revealed the finding that there was only an independent correlation between BMI and increased AP diameter/thoracic transverse diameter ratio (p < 0.001). COPD patients compared to normal controls showed an increase in thoracic AP diameter.

Robert S. Eisinger and Hans K. Pilegaard (2018)[118] focused on the development of embryonic and vertebral abnormalities in a review study with graphic images using imaging techniques. The purpose of this review article was to introduce the reader to clinical conclusions regarding normal and abnormal spinal anatomy, embryology, imaging, and different spinal deformities resulting from abnormal embryological development. The spine is subject to various developmental structural abnormalities. Vertebral deformities can result from defects in vertebral development, union, separation, and developmental differences. These malformations range from simple to complex structural and neurological problems. These abnormalities may be mistaken for acute trauma and may predispose to myelopathy. To accurately interpret spinal abnormalities in images, clinicians need to know normal and abnormal spinal morphology according to age and the clinical outcome of each entity. This knowledge increases diagnostic confidence in high-stress environments and problematic clinical conditions.

Samuel Y. Ash and Alejandro A. Diaz (2017)[119] conducted a review analysis to examine the strongest associations between imaging studies and severe asthma and their association with clinical outcomes. Prominent results include the use of CT imaging biomarkers to generate a population of asthma patients and the use of cross-registration to match the CT images of airway to regional differences in ventilation on MRI. Moreover, a temporary study published that certain ventilatory abnormalities discovered by MRI in an asthma patient were sporadic and others refractory, both associated with decreased lung function. The role of imaging in severe asthma is currently primarily to rule out comorbidities or other conditions, or to assess impairment in the setting of acute decompensation. The rapidly growing literature using CT and MRI recommends their usefulness in the long-term management of this disease.

Siti W Mohd-Zin, Ahmed I Marwan, Mohamad K Abou Chaar et al. (2017)[120] explored the correlations by both surgeons and embryologists and finally contrasted them with a prominent animal model, the mouse. In this review article they proposed and reported the spina bifida and phenotypes observed in humans. However, the understanding of spina bifida is presently restricted to the annotations made in mouse models, which redirect complete or targeted knockouts of genes, which bother the whole gene(s) without taking into account the matter of haploinsufficiency, which is most protuberant in the human spina bifida condition. The need to study spina bifida in all its morphologies, both aperta and occulta, is a complementary analysis of surviving human spina bifida, and measurements of erosion resulting from neural tube defects in the tail were identified in spina bifida. It was concluded to be more common than by the level of lesions in both mice and humans.

Emily M. DeBoer, David R. Spielberg, Alan S. Brody, et al. (2017)[121] identified a major study that could serve as a potential clinical tool for patients with asthma and other types of lung disease. A design-based analysis was performed to evaluate the best imaging modalities and some of the emerging innovations in lung imaging. Pulmonary magnetic resonance imaging, chest radiography, bronchoscopy, computed tomography, and electrical impedance tomography were some of the available imaging modalities. Studies associated with clinical and translational trials are increasing the ability of lung imaging to reveal patient profiles, prognosis, and response totherapy.

Irfaan Abid and MennatAllah M. Ewais et al. (2017)[122] observed that osteopathic medicine may focus on the musculoskeletal system and the fact that the anatomy of chest wall stimulation function is an important parameter. A common congenital deformity of the chest wall is known to be the sternum being pushed back with a funnel chest, and tachycardia, chest wall pain, and exercise intolerance are some of the common problems. There has been some debate among people regarding funnel chest, its medical consequences, diagnosis, and its treatment options. Funnel chest is known to be a common chest wall deformity that has become well known. Note that the osteopathic practitioner should also be aware of the structure of the chest wall. Chest wall structure can affect a variety of patient symptoms and functions. In certain circumstances, musculoskeletal and heart problems can have serious medical consequences. In mild cases, osteopathic maneuvers can provide clinically significant relief of chest wall pain.

S. Singh and S. B. Greenberg (2017)[123] conducted a study with the primary aim of investigating a new finding in patients with funnel chest, hyperinflation of the anterior base of the left lower lobe [LABSH]. A secondary aim was to know the association between the frequency of breakthrough findings and the severity of funnel chest. This study included 52 children who underwent preoperative computed tomography (CT) to assess funnel chest. 137 unrelated CT scans of children were included as a control group containing 50 children. Patients in both groups ranged in age from 12 to 20 years. Haller index was also assessed in all patients. LABSH was determined by visual inspection of lung windows. The difference [ΔHU] in mean radiodensity measurements in the left and right anterior fundus regions of interest was calculated. Spirometry was performed on 44 patients, and the results were related to the severity of the Haller Index and her presence of LABSH. Echocardiographic reports of 50 children in the funnel-chest group were also provided. In LABSH the Haller index was > 4.0 ($p = 0.001$). ΔHU in LABSH patients was also significant ($p<0.0001$). There was also a significant association of LABSH with the funnel-chest group compared with the control group. There was no significant association between children's mean Haller index differences and normal spirometry ($p=0.9899$) compared to children with obstructive disease (4.5, SD = 1). Two echocardiograms also showed mild cardiac compression. It was concluded that LABSH is a new indication associated with funnel chest. This suggests partial bronchial compression caused by chest deformation resulting in partial air entrapment.

Anda Maria Aignătoaei et al. (2017)[124] conducted a survey of 11538 chest radiographs and 315 CT imaging examinations in 2014–2015. This case series included three subjects with bilateral cervical ribs and one case each with Bifida and fused ribs. It was observed that all patients revealed rib abnormalities within the brackets. Various malformations of the ribs were detected on chest radiographs, except for non-union of ribs identified on CT scans. Differential diagnosis of cervical ribs, C7 abnormalities, and first rib and other rib malformations to distinguish from neoplastic, traumatic, or inflammatory lesions of the chest wall, lungs, and pleura. Therefore, given the general knowledge of rib development, the three most common types of rib malformations have been proposed: (I) numerical aberrations; (II) segmentation error, (III) re-segmentation anomaly. It was therefore concluded that it is important for radiologists to familiarize themselves with the imaging features of rib abnormalities, as these abnormalities can be misinterpreted as lesions with different meanings.

Nivedha Viswanathan, and Ayesha Parveen et al. (2017)[125] reviewed case reports discussing irregular fusion of vertebrae leading to a clinical condition called blocked vertebrae or vertebral synostosis. Of all known vertebral fusions, thoracic fusions were shown to be infrequent and relatively rare. Defects in vertebroplasty can cause neurological signs and symptoms. Possible reasons for the fusion of the thoracic spine may be the congenital spinal deformity DISH (diffuse idiopathic hyper-skeletal hyperostosis) and other rheumatic degenerative diseases or tuberculosis. This can lead to a wide range of problems affecting various bodily functions. This can also represent a variety of clinical signs, such as abnormal curvature of the spinescoliosis, and kyphosis.

Rajkumar P et al. (2017)[126] demonstrated knowledge gaps and ultimately developed a database that can conduct long-term cohort studies on the evolution of phenotypic and genotypic heterogeneity in COPD patients in India. The primary objective of this study was to estimate the incidence of COPD among adults aged 25 years and older by sex in India. A secondary goal was to identify risk factors for COPD and important complications such as asthma and tuberculosis sequelae. Therefore, a cross-sectional study was conducted in the suburbs of Chennai and Shillong cities representing the southern and northeastern parts of India. Data were collected on sociodemographic variables, economic characteristics, risk factors for COPD, and comorbidities. The Global Obstructive Lung Disease (GOLD) and Global Asthma Initiative (GINA) definitions were used to diagnose COPD and asthma. This study helped to clearly define COPD phenotype and genotype in follow-up studies

James E. Archer, Adrian Gardner, Fiona Berryman, et al. (2016)[77] In this article, the Haller index was measured by chest width (transverse diameter) and height (anteroposterior) at 10 levels of the thoracic vertebra from T3 to T12 of patients between 8 to 18 years of age. Used it to show the shape of chest. This varies by vertebral level, so the greatest ratio is found at the most cranial level of the thorax. With increasing age, the shape of the thorax changes, and most of the cranial vertebrae that exhibit a greater Haller index in the mid-thoracic region remain unchanged. They also observed that subtle changes were visualized in the more caudal vertebral level. These analyzes suggest that the "one size fits all" rule regarding chest width-to-depth ratios is not appropriate for all stages and levels of the chest and should serve as a basis for the normal range of width-to-height ratios.

Robert E. Kelly Jr., Robert J. Obermeyer, and Donald Nuss (2016)[127] examined patients with funnel chest using a variety of methods at various centers but found fatigability, and ataxia No symptoms of tolerance, pulmonary function, or shortness of breath were observed. Several imaging tests with plethysmography, spirometry, exercise stress testing, and ocular and electronic plethysmography has been used to assess pulmonary function and response to surgery in patients with funnel chests. Not all patients with funnel chest had subnormal static pulmonary function tests, but some had above-average values. However, in over 1500 adult and pediatric surgical patients, the postoperative curve was approximately 1 standard deviation. Previous experiments have shown that patients with anatomically severe pulmonary embolisms are susceptible to reduced PFT. This mechanism is visualized by electronic ocular plethysmography to confirm that the recessed area of the chest does not move during respiration. PE is also associated with static lung function decline in most patients. Correction by Nus surgery results in improved chest wall motion; this up-regulation of chest bellows action was complemented by improved pulmonary function tests.

Guglielmo M, Riccardo I, Eliseo P et. Al. (2016)[128] reported a case study of a 26-year-old Caucasian man admitted to the emergency department after a motor vehicle accident. On admission, the patient was hemodynamically stable (BP: 130 systolic blood pressure ≥75 mm Hg diastolic, heart rate: 84 BPM), dyspnea (SpO2: 80% pO2), facial swelling, and headache, no neck pain, no obvious skin deformities or subcutaneous masses. Cerebral hemorrhages were scanned with CT according to standard protocols. Neck pain was examined via X-ray of the cervical spine in anterior-posterior and lateral projections. Although no evidence of pneumothorax was found, minimal bilateral pulmonary contusion confirmed a lack of fusion of the T1 spinous processes, and the examination was repeated according to the specific CT protocol for cervical spine studies. Patients were monitored for 3 days and treated with anti-inflammatory drugs. No surgical intervention was required in this case.

Fizan Abdullah and Jamie Harris (2016)[129] highlighted the severity, impact of anesthesia, and outcome of repair surgery for PE. PE is the most common congenital thoracic wall defect, affecting men five times more often than women. The most common symptom is chest compression. Part of the evaluation is serial measurements of chest deformity. Complementary cardiorespiratory testing should be performed, including arrhythmia and pulmonary function testing. The Haller Index was calculated from CT scans and was considered severe if the score was greater than 3.2. The main indicator of repair is a decline in cardiopulmonary capacity. Surgical treatment should be performed after the child's growth spurt. The Nuss procedure is usually the surgical repair of choice. Ravitch repair is used for more complex or asymmetric malformations.

Mac SM et al. (2016)[130] During the full inspiratory phase, they used plain low-dose chest CT. The patient was scanned from the apex of the lung to the coastal recess using tomography parameters of 120 kV and 35 mA. Appropriate fenestrations are specifically designed to highlight the skeleton, costal cartilage and muscle, and skin. PE and carinatum are the most common congenital diseases affecting the chest wall at birth or in early childhood. They showed typical radiographic and CT presentations of these conditions. Cross-sectional CT imaging using low-dose techniques is reserved at her institution for patients undergoing corrective surgery.

Jin Young-Jeong, Jun Hyung-an, Sang-Yong Kim, et al. (2015)[131] evaluated the difference in pulmonary function before and after Nuss surgery and preoperative morphology. He conducted retrospective review of 18 male patients who underwent Nuss surgery. There were 9 patients with symmetrical and asymmetrical morphology, respectively. Patients younger than 18 years with no history of respiratory disease were recruited for the study. Lung function was assessed 2 weeks before surgery and 4-6 months after surgery. Asymmetric patients had lower preoperative vital capacity (VC) ($p=0.0308$), inspiratory capacity (IC) ($p=0.0373$), and total vital capacity (TLC) ($p=0.018$). For both types, forced expiratory volume in 1 second (FEV1), forced vital capacity (FVC), and VC was significantly decreased postoperatively compared to baseline (all, $p<0.01$). The asymmetric type showed postoperative decreases in peak expiratory flow (PEF) ($p = 0.0391$) and IC ($p = 0.0084$). The RV/TLC ratio ($p = 0.0025$) and residual volume (RV) ($p = 0.0092$) are significantly increased in the asymmetric type, but only the postoperative PEF values were significantly different between the two types ($p = 0.0151$). Preoperative lung volume was worse in asymmetric cases, postoperative lung function was worse, and PEF was significantly lower compared with symmetric cases. Asymmetric types of PE require closer assessment of preoperative and postoperative pulmonary function until the pectoral muscle is removed.

Mohd. Nazeer et al. (2015)[132] conducted a study aimed at presenting cases of thoracic spinal fusion. In this work, the clinical and developmental importance of fused vertebrae was discussed. In this literature, the incidence of fused thoracic vertebrae ranged from 1.6% to 4.16%. Unsuitable vertebral fusion causes vertebral synostosis, block vertebrae or spinal fusion. The neck is most commonly affected, followed by the chest and lower back. During vertebral column development, during the fourth week of life in the womb, the sclerotome portion of the somites migrate around the notochord and neural tube and become segmented. Errors in this process lead to spinal abnormalities, causing neurological signs and symptoms. It is caused by trauma, tuberculosis, and juvenile rheumatoid arthritis. This leads to various problems that disrupt various body systems.

Dharmraj Meena reported a case study (2015)[133] **found a**bout 50-year-old man in the government's Diagnostic Radiology and Advanced Imaging Division. Kota Medical College, Rajasthan. Only 15 ribs were mediated by him. On the left side, the absence of the 1st, 2nd, 4th, 7th, and 11th ribs were observed. On the right side, the 1st, 3rd, 8th, and 10th were missing. He also noted the anterior bifurcation of these missing ribs at the end of the left sixth rib. Right 5th, 6th, and bilateral hypoplastic 12th ribs fused posteriorly. He also revealed that congenital rib anomalies were associated with thoracolumbar blocks, hemivertebrae bones, and small sternum.

Patrick M. Poston et al. (2014)[134] conducted a study examining appropriate criteria for advocating repairs based on modified indices, incorporating the criteria established by the Haller index. A record of 75 PE patients who underwent CT scans was compiled. Associations between all Haller indices and corrected indices were evaluated and were clear for those with standard and non-standard chest dimensions as measured by the ideal thoracic index. An uncertain association was observed between the Haller index and the corrected index ($r = 0.79$; $p < 0.0001$). The Haller index and the corrected index showed a stronger correlation after excluding patients who deviated from ideal chest measurements ($r = 0.86$; $p < 0.0001$). The Correction Index provides an accurate assessment of disease severity and, by measuring it, re-directs the extent of possible surgical repair. The Haller index correlates well with the corrected index in pectoral patients with standard chest wall measurements but disagrees with nonstandard chests. This value is consistent with the long-accepted standard (Haller index ≥ 3.25), and this index remains accurate even with non-standard chest morphologies, making surgical repair of PE with a modified index of 28% or greater possible. It is recommended in the nonstandard chest also.

Selvi Asker, Muntecep Asker, and Bulent Ozbay (2014)[135] aimed to assess bronchial thickness by high-resolution chest CT in patients with mild intermittent asthma compared with healthy controls was carried out. In this study, 37 of outpatients with mild intermittent asthma and 13 of healthy controls were recruited into this study. Subject and control groups showed similar pulmonary function test results and total and partial T/D values. Mean ±SD subsegment WA% values above 2 cm above the inferior pulmonary vein and diaphragm were significantly higher in subjects than in controls. Results showed increased bronchial wall thickness in small airways in patients with mild intermittent asthma, regardless of asthma duration. This may indicate that effective anti-inflammatory or bronchodilator therapy should also be given to the small airways in the early stages of the disease.

Dhana Lakshmi V, Arun Kumar K.R., Satish Kumar S.et. Al. (2014)[136] describes a case of normal dissection of a bifurcated rib at the Institute of Anatomy, Madras Medical University. Several bifurcated ribs present in male cadavers showed various rib differences. This is a relatively rare condition and affects 0.15% to 3.4% of the population. Gorlin-Goltz syndrome is represented by a bipartite rib. Surgeons, physicians, and radiologists would benefit, according to the authors.

Sverzellati N, Colombi D, Randi G, Pavarani A, Silva M, et al. (2013)[137] conducted a study evaluating the effects of anthropometric and COPD-related thoracic changes in adult smokers. Chest diameter and area of 816 smokers with and without COPD were assessed at three levels of anatomy using CT. Gender was observed to have the greatest effect on rib cage diameter and area. Emphysema % was responsible for the enlargement of upper and middle CT regions. Reduced FVC% to shrink the lower chest. This study showed that simple CT measurements are sufficient to predict the variability of chest morphometry and also highlight the association between chest morphometry and emphysema.

Bastir M, García-Martínez D, Lechais W, Barrie A, Coquerel M, et al. (2013)[138] used geometric morphometry applied to extensive landmark data from the thorax to determine changes in growth. They conducted a study that reexamined the digitized 402 (half) landmarks in 3D reconstructions of the thoracic cage to assess changes in growth in 27 computed tomography-scanned modern humans representing neonatal to male and female adults.Their analysis reveals a curved ontogenetic trajectory resulting from different ontogenetic growth allometry of the upper and lower thoracic units. Adult thoracic morphology arrives later than predicted due to variable changes in different anatomical regions at different ontogenetic stages. In addition to the significant increase in anteroposterior dimensions, there is an increase in medial-lateral dimensions of the upper chest compared to the lower chest. This transforms the pyramid-shaped infant chest into a barrel-shaped adult. Rib descent is created by a complex change in 3D curvature. Developmental differences between the upper and lower thorax are related to differences in the timing and rate of maturation of the respiratory and digestive systems, the spine, and the musculoskeletal system. Their results are relevant to understanding how changes in the relative growth rates of these systems and structures affected the development and evolution of modern human morphology.

Michele Gaeta, Fabio Minutri, Giuseppe Sirvino and others. (2013)[139] conducted a study to find annihilating bronchiolitis, which can cause abnormal breathing. They underscore the criteria for this method in patients with clinical and functional symptoms of bronchiolar obstruction who underwent a CT scan despite normal inspiratory CT. Expiratory CT scans are often used in combination with conventional inspiratory CT scans to detect air entrapment, a sign of mild airway obstruction. Air pockets may be caused by early-stage minor airway disease in otherwise healthy patients.

Rita Priori, Andrea Aliverti, André L. Albuquerque et al. (2013)[140] observed that various positions from sitting to lying affected thoracic asynchrony through changes in respiratory muscle conformation. Twenty-three of their severe COPD patients and 12 age-matched healthy controls were taken under study. Thoracic asynchrony was observed in a sitting position in 13 of COPD patients. Thoracic asynchrony disappeared in the supine position. In COPD, there was strong asynchrony in the supine position, but synchrony between the upper ribcage and abdomen when seated. These findings were supported by the relationship between diaphragmatic apposition zone variation and chest wall compartment volume differences. Supine position means that the thoracic paradox disappears and diaphragmatic mechanics was restored in this position. It was concluded that postural differences in COPD patients lead to significant changes in chest wall behavior.

Mousa K, Jayakrishnan B, Abdulaziz M, and others. (2012)[141] attempted to correlate these abnormalities with clinical and pulmonary function test (PFT) data. Kuwaiti underwent HRCT for 6 months after initial evaluation and PFT of a patient with stable asthma attending Mubarak Al Kabeer Hospital. Of the 28 patients, 16 had moderate (57.1%), 6 (21.4%) mild, and 6 (21.4%) severe persistent asthma. Bronchiectasis (28.6%), squamous atelectasis (21.4%), bronchial wall thickening (57.1%), mucoid obstruction (17.9%), air entrapment (78.6%), and mosaic opacity (10.7%) was evaluated. The most common was bronchial wall thickening (p=0.044) and bronchiectasis in men (p=0.063). Ten (35.7%) patients had mild, 9 (32.1%) moderate, and 3 had (10.7%) severe air entrapment. Hounsfield units between expiratory and inspiratory cuts (air trapping) correlated with predicted FEV1 percentages, right middle (r = 0.15; p = 0.53), upper right (r = 0.25; p = 0.30), Left middle (r=-0.04; p=0.60) and upper left (r=0.20; p=0.41) lower left zone (r=-0)13; p=0.58) and bottom right (r=0.04; p=0.86). No significant correlation swaths were found when correlating with the FEF 25–75 proportion predicted as above. Air trapping was compared with nationality (p=1,000), symptom severity (p=0.581), gender (p=0.640), and disease duration (p=1,000). However, they found that the presence of air entrapment was not related to disease duration or severity, or FEV1.

Ti-Hei Wu, Tsai-Wang Huang, Hsian-Hsu, etc. (2012)[142] conducted a study to assess whether chest radiographs can provide valuable information for the assessment of

PE before and after Nuss repair in adults. In this study, 154 adults with a mean age of 24.0±5.0 years who underwent Nuss repair for LE with a mean follow-up of 42 months were recruited into this retrospective study. Sixty-two patients were followed after PE removal, with a median follow-up of 13 months. The preoperative and postoperative imaging methods were compared. The mean preoperative Haller index was 4.61 ± 1.58 (range: 2.6-11.9) and the chest radiograph was 3.82±1.17 (range: 20-10.2). The Pearson correlation coefficient between the two variables was 0.757. The mean postoperative Haller index showed a statistically significant improvement compared with the preoperative index (P < 0.001). The mean sternum-spine (SV) distance recorded preoperatively and postoperatively (at least 6 months postoperatively) was 7.67 ± 1.89 cm (range 2.5–12.9 cm) and 9.89 ± 1.80 cm on lateral chest radiograph. (Range 4.6–15.0 cm).), statistically significant improvement (P< 0)001). After bar removal, the patient's mean SV distance was 9.25 ± 2.14 cm, a statistically significant improvement compared to preoperative values (P<0.001). Therefore; it was concluded that the Haller index and CT show a strong correlation. Therefore, chest radiographs can be used as a surrogate tool for the preoperative evaluation of PE. Nuss surgery significantly improved the perceived SV (stroke volume) distance on chest radiographs.

Marie T. Izatt et al. (2012)[143] conducted a prospective study on 30 Renke-1 patients with adolescent idiopathic scoliosis (AIS) treated with elective thoracoscopic spinal fusion (TASF). At least 24 months; participants underwent an ethically approved low-dose CT scan postoperatively in addition to standard postoperative care. Changes in the sagittal contour of the lying CT were equated with the patient's standing radiographs and previous studies. Inter-observer disagreement was assessed to assess whether patient groups with hyperkyphosis and normal kyphosis responded adversely to the anterior thoracoscopic methodology. At assessment, CT measurements of kyphosis increased by 12.3 degrees and lordosis by 7.0 degrees. Restoring the sagittal profile is an important goal in scoliosis surgery, but consistent measurements with radiographs suffer from poor endplate clarity. TASF was observed to significantly improve lumbar lordosis and thoracic kyphosis and preserve the configuration of the proximal and distal junctions in patients with thoracic AIS. They found that supine CT for sagittal Cobb measurements allowed for excellent clarity of the endplates, and supine CT and upright radiographs had linear measurements. This study also showed that anterior correction was preferred over posterior correction to address the sagittal portion of instrumented and non-instrumented segments after progressive surgical improvement of Renke-1 idiopathic scoliosis. has also been shown to be effective.

Lawson M. Louise, Robert B. Mellins, James F. Paulson et al. (2011)[144] examined whether the severity of pulmonary embolism results in decreased lung function and whether the nature of the function is limiting. They conducted a study to determine whether the Lung volumes of 310 patients and 218 patients aged 6 to 21 years were assessed with preoperative spirometry data at 11 centers in North America. The effect of deformity severity (based on the Haller Index) on pulmonary function was also assessed. The proportion of patients with findings of abnormal FVC (FEV (1)), forced expiratory flow from 25% to 75% of exhalation, and total lung volume supplemented with cumulative Haller Index score. Less than 2% of patients were seen with obstructive pattern and 14.5% with restrictive pattern and <80% predicted. A patient with a Haller index of 7 is more than four times more likely than a patient with a Haller index of 4 to have an FVC ≤80% and four times more likely to have a restrictive pulmonary pattern. Among patients presenting with surgical repair of PE, those with more severe strains are much more likely to have decreased pulmonary function with an obstructive pulmonary pattern.

Wojciech, Robert Sitnik MW, etc. (2009)[145] provided an automated assessment method for PE severity based on marker-free 3D optical profilometry. A four-dimensional measurement system based on structured light projection was set up to record the shape of the patient's body surface. System settings were detailed, characteristic measurement parameters were presented, and automated data analysis paths described. Its main stages are normalizing the orientation of the torso model, slicing the model, analyzing the shape of each slice, selecting suitable slices for the patient's evaluation of PE, and calculating shape parameters. A developed new shape parameter I(3ds) that showed a high correlation with the CT-Haller index was widely discarded to assess PE.

Michel M, Mathieu L, Stephanie R, Hugues B, etc. (2009)[146] conducted a retrospective study investigating and comparing CT bronchial measurements in asthma patients and healthy controls and their correlation with bronchial morphometric parameters. In their CT and pulmonary function tests of 27 subjects were divided into two groups of 15 asthmatics and 12 healthy subjects. Endobronchial biopsies were also performed in 11 subjects. Lumen area (LA), bronchial cross-sectional area (WA), wall thickness (WT), WA/LA ratio, total area (TA), and WA/TA ratio were calculated using validated software. Patients with asthma had significantly lower TA, LA, WA-WA/TA ratios, and higher WA/LA ratios compared with fourth-generation healthy controls. Slope and maximal regional slope of WA/TA and WA/LA ratios and subepithelial thickness (r = 0.67–0.69, p = 0.019–0.023), bronchial obstruction (r = 0.46–0.58, p = 0.001) There was an association between functional data including–0.025), subepithelial area (r=081, P=0.002), smooth muscle layer area (r=0.75, P=0.007), and bronchial wall infiltration by inflammatory cells (r=0.67-0.86, P=0.049-0.003). Axial reconstruction and orthogonal measurements, as well as airways enabled by three-dimensional segmentation methods, reveal significant variation in bronchial morphometrics, predict airway restriction in asthma, and provide a noninvasive measure of airway remodeling, that can play a role in the Non-invasive measurement of airway remodeling.

Aliverti A et al. (2009)[147] studied 20 patients and their 10 healthy controls at rest and during symptom-limited progressive physical activity. Used control data to define the normal range of motion. Photoelectron plethysmography, quantification of lung and abdominal thoracic volumes, phase shifts, and a fraction of inspiratory time in which the thoracic compartments progress in opposite directions were performed. Eight patients showed reduced thoracic inspiratory paradox at rest (P+) and 12 patients did not (P-). This was not associated with resting lung function or exercise tolerance. Total end-expiratory chest wall volume (EEVcw) increased immediately after exercise initiation in P+ patients, but increased later during exercise in P patients. This EEVcw difference was primarily due to the increased end-expiratory pulmonary-thoracic volume in P+ patients. During exercise, dyspnea was observed to increase equally in both groups, whereas leg effort increased more significantly in patients without paradox, which was associated with early-onset chest wall hyperinflation and prime dyspnea at the end of the exercise.

In the absence of paradoxes, leg fatigue becomes an additional compelling exercise to limit symptoms.

Fokin AA, Steuerwald NM et. al. (2009)[148] In their investigation, there was a diverse group of congenital malformations of the chest, which manifest as anterior chest wall defects and, depending on the severity and associated irregularities, may have cardiopulmonary implications. Pectus excavatum (PE) is a common precordial malformation, with a depressed sternum with consistent left displacement and cardiac rotation. Pectus carinatum (PC), the second most common, presents with variable chest wall protuberance and highly variable clinical indices. The abnormal elongation of the costal cartilages served as the reason for these conditions is thought to be. Collagen disorder is concerned with genetic and histologic analysis for the pectus excavatum and carinatum. A sternal fissure is a rare birth defect caused by the lack of fusion of the sternum halves, leaving the heart vulnerable to come out of the thoracic cavity and requiring early surgical intervention. The abnormal opening in the sternum bone is related with cardiac anomaly is known as ectopia cordis.

Marcus PK, Peder GN, Margaret WL and et. al. (2007)[149] reviewed all high-resolution chest CT studies for a cohort of patients with primary ciliary dyskinesia (PCD) to identify relevant patterns of high-resolution CT changes. These were available for 45 PCD patients from 42 families with varying ages and disease severity. Images were analyzed for severity and distribution of peribronchial thickening, bronchiectasis, mucus obstruction, and other findings. A bronchiectasis severity score was also calculated. CT findings were associated with phenotypic findings including ultrastructural defects, FEV1, site type, ciliary nasal nitric oxide levels, and airway microbiological findings. Twenty-nine adults and their 16 children participated in the study, accounting for 58% of patients. Heterotaxy (18%) and total internal inversion (38%) were found in 56% of patients and 9% PE was known. High-resolution CT scans in 56% of all adult and pediatric patients showed bronchiectasis, primarily in the middle and lower lobes. The right middle lobe was most commonly affected. Bronchiectasis severity was associated with older age and decreased lung function. High-resolution CT showed that pulmonary disease associated with PCD mainly affected the middle and lower lobes of the lung. In adults, high-resolution CT findings negative for bronchiectasis served to rule out the diagnosis of PCD. The association between disease severity on high-resolution CT and patient phenotype provides further insight into the diversity and natural history of PCD.

Dawn E. Jaroszewski, Eric W. Fonkalsrud, et. al. (2007)[150] conducted a retrospective review of a patient aged 19 years or older who underwent surgical correction of PE in the hospital setting. Patients presented with tachypnea with dyspnea, decreased endurance, chest pain (69%), mild exertion (99%), and tachycardia (94%). In all patients, symptoms worsened during adolescence and increased in severity in adulthood. Symptomatic improvement was observed in all patients within 4 months of surgical repair. The average length of hospital stay was 2.9 days. Postoperative complications such as pneumothorax (n = 4), pleural effusion (n = 8), and pericarditis (n = 2) were found. Eight patients had small focal protrusions. 4 patients had undergone repair of mild recurrent deformities. No deaths were seen. 98% showed significant improvement in exercise tolerance and rated the postoperative outcome as good or excellent. Uncorrected pectus deformities which persist during childhood worsen with advancing age. Repair can be done in adults with those having low morbidity, short hospital stays, and substantial development in physiologic function.

Deborah M. Kado, Katherine Prenovost, and Carolyn Crandall (2007)[151] A recent study estimated the prevalence of hyperkyphosis in the elderly to be between 20% and 40%. Its cause and meaning are not fully implied. Some doctors believe that fractures are the cause of hyperkyphosis and that treatment strategies should focus solely on diagnosing and treating osteoporosis. Recent researches validate that many older adults are mostly affected by hyperkyphosis and have no vertebral fractures. Hyperkyphosis might be independently associated with an enhanced risk for poor health outcomes, including diminished physical function competences, impaired pulmonary function and upcoming fractures. With the advancing age research is needed for deeper understanding of the causes, consequences and treatment.

PA de Jong et al. (2005)[152] conducted a study aimed at investigating the use of CT in studying airway anatomy and function in health and disease. Airway remodeling, or marked changes in airway structure leading to airflow limitation in various chronic lung diseases. Histology has provided a startling vision for the structure of airway wall remodeling, but it is an invasive procedure that is unsuitable for longitudinal studies of airway wall dimensions. Improvements in CT technology have enabled the quantification of airway wall measurements, making them suitable for noninvasively investigating the etiology of airway wall remodeling and evaluating potential therapeutics.

V. Ne`ve, F. Girard, A. Flahault, M. Boule and et al (2002)[153] conducted a study with an aim to standardize the reference values for the lung transfer factor for carbon monoxide (TL, CO) and lung volumes and thoracic volume index (TVI) in adolescent population and an index of pubertal stage in order to rule out the variation in adolescents' growth pattern. Time since menarche, total lung capacity and TL, functional residual capacity measured using the helium-dilution technique, TVI, pubertal stage by Tanner scale (PST), vital capacity, CO measured using a steady-state method were identified in 51 males and 52 females. In male it was observed that with increasing age weight, height, adolescents, TVI, TL, lung volumes and CO increased. It was not seen in female's adolescents. In males, it was observed that the TVI was an independent variable and had the best correlation with pulmonary volumes and in females, height was an independent variable best correlated with the same. In both the cases, in association with height in males, the variable that best interrelated with TL, CO was PST. This cross-sectional study aided in the prediction equations for lung transfer factor for CO and lung volumes and in relation to pubertal stage and thoracic volume index. It was also seen that in adolescent males, thoracic and lung development occurred during and until the end of puberty. Contrariwise, in adolescent females, lung development is almost over following menarche.

Ronald B.J. Glass et al. (2002)[154] Review highlights include: Congenital and acquired manifestations of this disease can be found in children's ribs. Normal variants are not clinically significant. It is rarely palpable on clinical examination or incidentally seen on chest x-rays. Signs of abnormality can be seen in the ribs as differences in size, number, shape, and calcification. These changes can be focal or generalized. Abnormalities found in the ribs may be the first sign of a previously unanticipated systemic disease. Ribs may provide important diagnostic clues in the scrutiny of patients with acquired metabolic diseases, congenital bone dysplasia, tumorigenesis, iatrogenic conditions, trauma (particularly child abuse), and infectious diseases. Routine assessment of the ribs with each chest radiograph is important to avoid overlooking important diagnostic information. Diagnostic data obtained from rib evaluation may help guide the radiological and laboratory investigations essential to complete the patient's diagnostic workup.

Adam R. Guttentag, Julia K. Salwen (1999)[155] Various pathological conditions and normal variants can be missed if the ribs are not carefully evaluated. The normal rib variant consists of fusion and bridging. intrathoracic and pelvic ribs; cervical, bifurcated ribs; and non-union of the first rib. Trauma-related lesions, which often occur in isolation, may alert the radiologist to other injuries. Cartilage-like lesions are found mainly near the anterior ends of the ribs. A calcification of the cartilage cap of the rib that represents a deformity or dilation is called an osteochondroma. The foci of bone destruction represent acute rib infections, whereas chronic infections involve bone sequestration and periosteal reactions. In most cases of pathological conditions, a subcostal indentation is often observed. Rib abnormalities can also occur in Paget's disease, hemoglobinopathy, fibrous dysplasia, and Langerhans cell histiocytosis. In most cases, radiographs are sufficient to diagnose normal variants and pathological rib conditions.

David S. Gierada et al. (1998)[156] conducted a study using magnetic resonance (MR) imaging to determine lung volume accuracy in patients with emphysema and found changes in chest dimensions after lung volume reduction surgery. Twenty-one patients with severe emphysema underwent rapid gradient echo respiratory arrest with complete inspiration and expiration, which was repeated in nine patients who underwent surgery. Postoperative variation in thoracic structure was calculated by measuring the lateral and anteroposterior lung dimensions and lung height before and after surgery. After surgery, chest dimensions decreased, and these types of decreases were greatest during expiration. Magnetic resonance measurements of lung volumes are similar to CT measurements but differ from plethysmography measurements. Variation in thoracic dimensions after lung volume reduction surgery is constant due to improved respiratory mechanics.

Donald Nuss, Robert E. Kellyr et al. (1998)[157] conducted a study evaluating the results of 10 years of experience with minimally invasive surgery without cartilage incision or PE resection. Between 1987 and 1996, 148 patients were examined for chest wall deformity. Fifty of the 127 patients suffering from PE were nominated for surgical correction. Eight elderly patients underwent Ravich surgery and 42 patients under the age of 15 were treated with minimally invasive techniques. Complications found included pneumothorax in 4 patients and required thoracotomy in 1 patient. Wound infection in 1 patient; steel rod displacement in 2 patients. (1) in three patients the bar was soft, (2) in one patient with Marfan syndrome the sternum was soft, (3) in one patient the bar was removed too early in the chest. Good and bad results were identified early because anomalies occurred. A minimally invasive procedure that does not require cartilage incision or excision was effective. By increasing the thickness of the steel bars at the same time and inserting two bars where needed, they had excellent long-term results. The upper limit is still being evaluated.

Janet E. Kuhlman et al. (1993)[158] studied the radiographic and CT manifestations of irregular air spaces in the lungs that develop in response to pulmonary disease. The main types of lung disease among them include pulmonary fibrosis, vascular embolism, adult respiratory distress syndrome, bronchiectasis, emphysema, air block disease, and rare diseases of the lung (tracheopharyngeal papillomatosis, Langerhans cell disease). histiocytosis and Klippel-Trenone syndrome). After reviewing conventional radiographs, CT scans, medical archives of 150 patients and 300 lung specimens, and high-resolution CT scans, the airspace formation process falls into five basic categories:
(1) lung architecture and retractile fibrosis remodeling (2) elastic fiber network disruption in the lung (3) bronchial dilatation (4) vascular occlusion or ischemic necrosis (5) multifactorial unknown mechanisms.

Haller JA Jr. and et. al. (1987)[159] reviewed a case report revealing that the thoracic lateral diameter divided by the anteroposterior diameter on a plain CT scan may be the pectus index. In the original report, all patients requiring surgical correction of PE had a pectus index (>3.25), whereas all matched normal controls had (<3.25). A plain CT scan helps evaluate PE surgery in children and adolescents individually.

Gerald T. Fong et al. (1980)[160] Chest radiographs of 316 healthy subjects using the Cobb method used to measure scoliosis for evaluation of thoracic kyphosis. Normal patients were those with no spinal or thoracic symptoms or chest radiographic abnormalities related to the thoracic spine. A total of 159 male and 157 female subjects aged 2 to 77 years were studied. The association between age/sex and kyphosis sorts the first-linear mathematical model. The results were also used to observe the predictable range of kyphosis in 'normal' patients of a given age and gender. The degree of kyphosis increased with age, and the proportion was higher in women than in men.

J.L. Ehrenhaft, Nicholas P. Rossi, and M.S. Lawrence (1966)[161] Developmental abnormalities and thoracic malformations have been major areas of concern for orthopedic surgeons and pediatricians. Growth plays a vital role, since the deformities can serve as an outcome of an inequity produced by the congenital abnormality itself or by forces, typically of muscular origin, operating on the same developmental problem. The progress of many closely connected bony and muscular systems, including spinal processes, vertebral column and ribs; sternum is obligatory for proper thoracic cage growth. Irregular development of the upper extremity or diaphragm muscles and intrinsic muscles of the chest wall may be abnormal. Numerous developmental disorders often co-exist, but even in their absence, one congenital anomaly can skew other closely related systems. The developing and growing thoracic spine and rib cage can be affected by the thoracic visceral disease. Scoliosis (congenital or acquired), if not stabilized during growth, can have a significant impact on the entire chest, leading to physiological abnormalities in the respiratory and circulatory systems. Surgical correction is essential for individuals because it can reduce the deviation and stop its secondary meaning before it becomes permanent once the growth stage is complete. Other minor malformations may not require treatment. Even if the symptoms are purely psychological or aesthetic, a systematic review should be conducted to see if the patient's appearance can be improved.

MATERIALS AND METHODS

MATERIALS

Simens computed tomography machine 64 slice, CT console, Attached desktop with CT machine, laptop, blank CD, camera, inbuilt PACS software with CT machine and 192 normal and 192 COPD patients.

STUDY DESIGN

Case-control

STUDY AREA

T.S. Misra Medical College and Hospital, Amausi, Lucknow, and private scanning centers of Lucknow city, Uttar Pradesh.

STUDY PERIOD

Two years (24 months)- one-year for forming study groups and data collection, six months for data analysis, and terminal six months for report writing and thesis submission. The study was started after the approval of the scientific committee and Institutional Ethical Committee of Nims University, Rajasthan, Jaipur, and T. S. Misra Medical College and Hospital, Amausi, Lucknow. The study began from January 2020 to December 2022 for a period of two years. The study was conducted in the association of Department of Anatomy and the Department of Radiodiagnosis of the T. S. Misra Medical College and National Institute of Medical Sciences & Research, Rajasthan, Jaipur. During this period four six monthly reports were submitted. In first six months from January 2020 to June 2020 study groups were formed, introduction and review literature was completed. In second six months from July 2020 to December 2020 data of normal group was collected and methodology was written. In third six months from January 2021 to June 2021 data of COPD group collected, first oral presentation for first publication was given in NATCON-68 and subsequently the first article published and submitted to the Ph.D. cell. In last (4th) six months' data analysis was done written work has completed. Second oral presentation was made in ACBI Pratapgarh later on second paper from the present study was published. Fourth progress report along with article and presentation certificate in consultation with guide and co-guide submitted in Ph.D. cell of the University on 16 December 2022.

SAMPLE SIZE

While calculating sample size things that were kept on the top-

1. Confidence level 95%
2. $Z_{1-n/2}$ is standard normal variate its value is 1.96 at 5% type I[st] error.
3. A confidence interval or margin means desired precision 'd' (5%, thisexplains permitted defectiveness – 5% i.e., mean lies between 95 &105).
4. The research design will be case-control.
5. Case and control sample ratio will be maintained to 1:1
6. 80% of 3.9 = 3.12, the difference considered to be clinically significant (James E. Archer et al, 2016)[11]

   Total sample size (N) = n_1 (case) + n_2 (control)

   Whereas $n_1 = n_2$

   Sample size was calculated on the basis of variation in Haller index in normal subjects and assuming the equality of variation in two groups under null hypothesis using the formula,

$$n = k\frac{\left(z_{\square} + z_{\square}\right)^2\left(\square_1^2 + \square_2^2\right)}{d^2}$$

Where $\sigma_1 = 2.85$, The SD of Haller index in normal subject.

$\sigma_2 = \sigma_1$, The SD of Haller index in case group under null hypothesis

d = 80% of mean maximum Haller index (=3.9), the difference considered to be clinically significant. k = 8, the design effect considering within groups of age and gender

Type I error α = 5% corresponding to 95% confidence level, Type II error β = 10% for detecting results with 90% power of study (Tale -2, value of Z_β = 1.84).

$$N = 8\times\frac{(1.96+1.84)^2\,(2.85^2 + 2.85^2)}{3.12^2} = 192.78$$

N = 192 + 192 = 384

Hence included a total of 384 patients taken under study.

INCLUSION CRITERIA

1. Indian residents were only enrolled
2. Both males and females were equally considered without chest deformity
3. Age limits under study were 18 years to 70 years
4. Random sampling method was adopted
5. Only those Computed Tomography thorax images were selected whose High-Resolution Computed Tomography findings were normal
6. Indoor and outdoor patients were equally considered
7. Males and females who had a history of Chronic Obstructive Pulmonary Disease
8. Patents whose High-Resolution Computed Tomography findings were suggestive of chronic bronchitis
9. Patents whose High-Resolution Computed Tomography findings were suggestive of emphysema
10. Patients who had given written informed consent in both groups

EXCLUSION CRITERIA

1. Patients who were operated upon a reconstruction of the thorax
2. Patients who were operated upon implanted Nuss Barr
3. Patients who had a history and operated upon open-heart surgery
4. Patients with implanted Pacemakers insitu
5. Patients with deformed chests due to trauma were excluded from the study
6. Patients with pregnancy in any trimester were excluded from the study.
7. Patients with diagnosed cases of pectus excavatum, pectus carinatum, and chest wall anomaly
8. Patients with scoliosis and kyphotic spine deformity
9. Patients with osteoporosis and spine surgery.
10. Patients whose High-Resolution Computed Tomography findings were suggestive of pleural effusion.
11. Patients whose High-Resolution Computed Tomography findings were suggestive of tuberculosis.
12. Patients whose High-Resolution Computed Tomography findings were suggestive of interstitial lung disease.
13. Post-infective sequel and post-covid-19 sequel.
14. Patients whose High-Resolution Computed Tomography findings were suggestive of Pneumonitis.
15. Patients whose High-Resolution Computed Tomography findings were suggestive of Sarcoidosis.
16. Patients whose High-Resolution Computed Tomography findings were suggestive of Bronchiectasis.
17. Patients whose High-Resolution Computed Tomography findings were suggestive of Malignancy and metastasis.
18. Patients whose High-Resolution Computed Tomography findings were suggestive of space-occupying lesions of the soft tissues of the lung parenchyma and mediastinum were excluded from the study.

METHODOLOGY

Institutional ethical clearances were obtained prior to the study's conduct at NIMS University Rajasthan, Jaipur, and T.S. Misra Medical College & Hospital, Amausi, Lucknow. Using inclusion and exclusion criteria, 384 patients were included who were willing to participate after providing written informed consent. The random sampling method was used. Patients who visited the departments of medicine, TB and chest (pulmonary medicine), paediatrics, and orthopaedics for chest-related issues were then rushed to TS Misra Medical College and Hospital's radiology department for X-ray chest and CT thorax. Both outdoor and indoor patient populations were equally considered. Similarly, samples were collected from the radiology department of the National Institute of Medical Sciences and Hospital, Rajasthan, Jaipur.

Separate study groups were formed. These groups were classified into two groups. The control group was Group I, and the case group is Group II. The sample population's "Group I-control group consists of those who have no significant findings on high-resolution computed tomography of the thorax. The sample population's Group-II case group includes those with significant findings on high-resolution computed tomography of the thorax for chronic obstructive pulmonary disease (bronchitis and emphysema), chronic bronchial asthma, and bronchiectasis. Previously, many other studies had been conducted by applying Haller`s methodology abroad, but few or no studies were seen in north India. Also, this study will be of its own kind because it is rare to see Haller`s Index measured at multiple vertebral levels, starting from T3 to T12 (10 levels), in chronic obstructive pulmonary disease patients. Most studies, whether conducted in India or abroad, involve only four to six levels for patients with chronic obstructive pulmonary disease.

All computed tomography thorax images were observed at 1.5 mm slice thickness in the bone window for measurement. Only good-quality CT thorax images were selected for observation and measurement. The desktop connected to the CT machine shows files of individual patients' achieved after imaging. These stored images can be seen in three dimensions and in three planes as well. The thorax image appears in the sagittal plane, the coronal plane as a whole, and as an axial image in the axial plane at a particular vertebral level that the observer is intended to see. The computed tomography thorax image in the sagittal plane was used as the base and guide for the measurement, and this image was used as a tool and template for the decision and determination of vertebral levels. As this plane delineates the sternum in front and the vertebral column behind, the sternum has two fusion sites: (i) the manubrium with the sternum's body at the manubriosternal joint, and (ii) the sternum's body with the xiphoid cartilage at the xifisternal joint. The vertebral column behind is usually seen from the cervical to the thoracic vertebrae, and in some images, the upper lumbar vertebrae may be present. The scaling tool can be used to move the desired vertebral level along the image from upwards to downwards or downwards to upwards. An axial image can be archived at a particular vertebral level. There are tools for measurement: curved, linear, and many others for measurement of volume and area. In the present study, a linear stool was selected for measuring linear distances, whether transverse or anteroposterior, in an axial image.

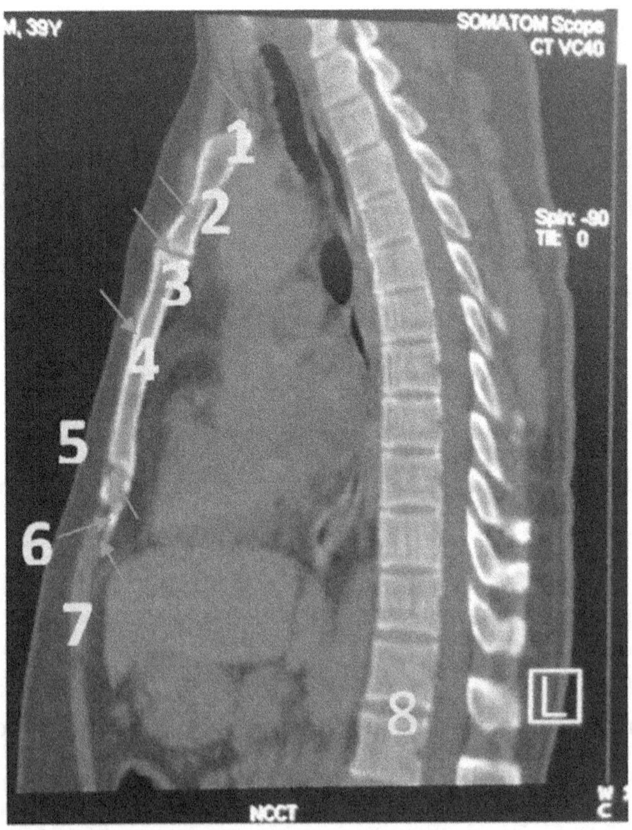

Figure 3.1 CT thorax image of a 39 years old male in sagittal plane showing

1. Upper end of sternum
2. Manubrium
3. Manubriosternal joint
4. Body of sternum
5. Xifisternal joint
6. xiphoid cartilage
7. Linea alba
8. Vertebral column

The measurement for each CT thorax in every patient started at thoracic vertebral level T3 because the superior border of the manubrium sterni (suprasternal notch) coincides with the lower border of the 2nd thoracic vertebra or intervertebral disc between thoracic vertebrae T2 and T3 posteriorly (Grays). Hence, the axial image of the CT thorax was achieved at the plane passing just above the lower border of the third thoracic vertebra. As there were sexual variations in the mentioned landmarks, Therefore, in case of any confusion, at a particular thoracic vertebral level, cross sectional anatomy was kept in mind, for which a standard atlases of CT thorax images in the axial plane were prepared based on Aquino et al 2001[162] for decision, comparison, and standardization.

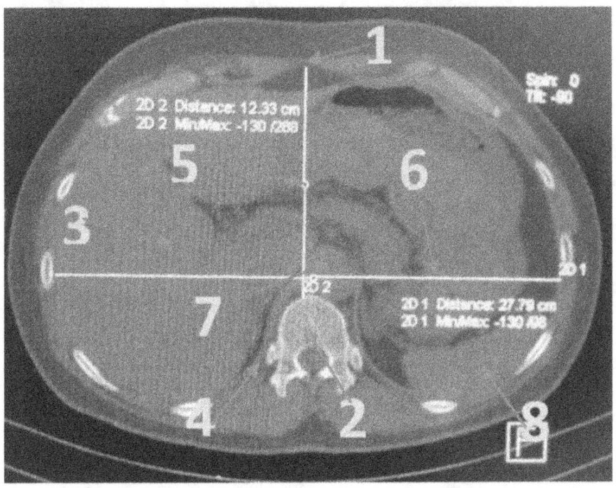

Figure 3.2 CT thorax image of a 39 years old male in axial plane at T11 showing

1. Anteriorly linea alba
2. T11 thoracic vertebra
3. Liver
4. Right Crus of diaphragm
5. Minimum internal anteroposterior diameter
6. Maximum internal Transverse Diameter
7. Thoracic aorta
8. Spleen

The Picture Archiving Communication System (PACS) built into the Siemens 64-slice CT machine was used to measure all distances. Only linear measurements were taken in digital soft copy images using the PACS linear measuring tool. The measurement of anteroposterior and transverse diameters began only at the T3 vertebral level. as there is no other identifiable fixed bony structure above T3 level anteriorly. The minimum internal anteroposterior diameter of the thoracic cavity was measured between the posterior surface of the sternum anteriorly and the anterior surface of the body of the respective thoracic vertebra posteriorly. At each vertebral level, the maximum transverse diameter was measured transversely between the internal surface of the right-side rib and the internal surface of the left side rib. Both the anteroposterior and transverse diameters are measured in such a way that they intersect each other at the right angle on the axial CT thorax image. In the same manner, measurements were performed from T3 to T10 thoracic vertebral levels.

Figure No. 3.3 – (a) (b) (c) (d) representing Steps of Measurement of transverse and anteroposterior diameter in axial plane

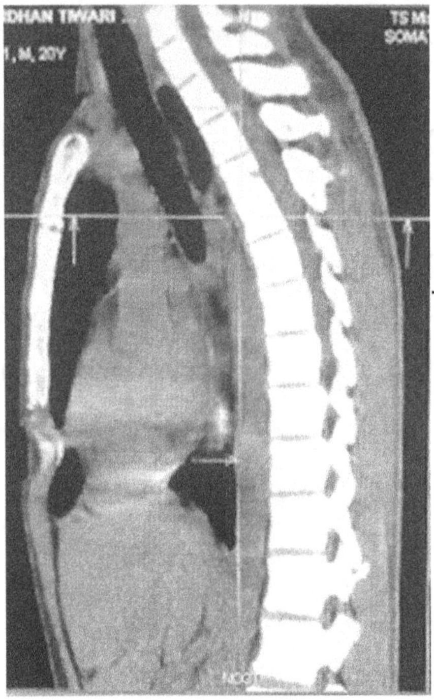

Figure No. 3.3 - (a) CT thorax in Mid sagittal plane showing Fixation of vertebral level

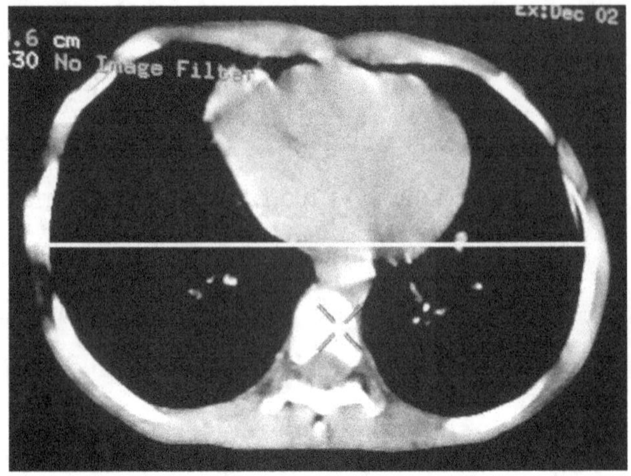

Figure No. 3.3 -(b) Measurement of Transverse Diameter

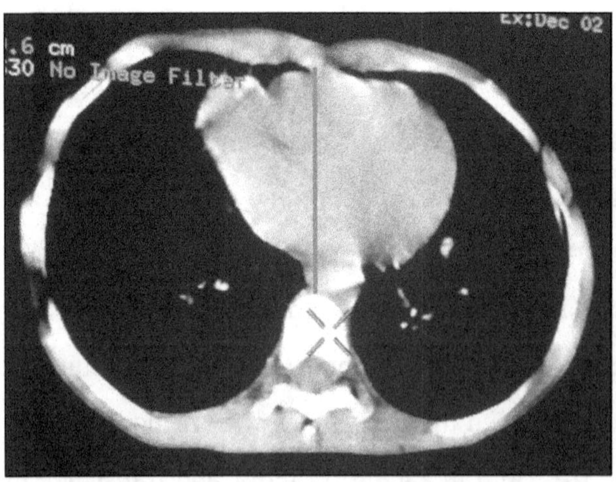

Figure No. 3.3 - (c) Measurement of Anteroposterion Diameter in axial plane

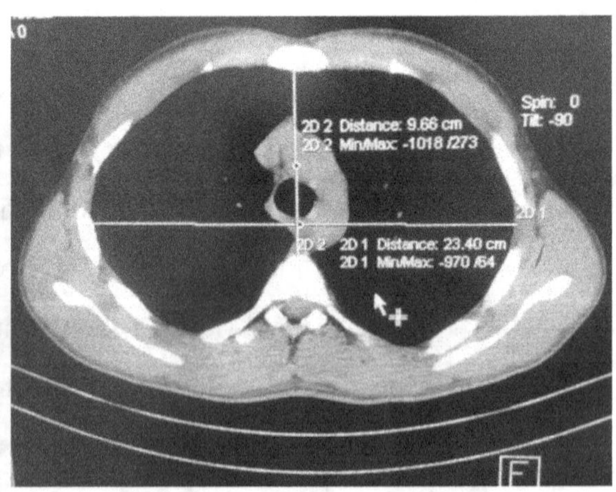

Figure No. 3.3 - (d) Transverse and anteroposterion Diameter measurement at right angle to each other

The anterior linea alba was used as a landmark below vertebral level T10. Therefore, below T10 level, the minimum internal diameter was between the linea alba anteriorly and the anterior surface of the body of the vertebra posteriorly. Linea alba lies aligned vertically in the plane of the sternum below the caudal end of the sternum. Measurements were made in the same fashion as previously. To avoid variations or discrepancies in measurement at each vertebral level and maintain uniformity, all the measurements were made just above the lower border of each vertebra.

Each and every individual has their own experience, perception, and interpretation; therefore, to avoid person-to-person measuring errors. The research scholar had taken training under the supervision of a co-supervisor who is faculty at T. S. Misra Medical College and Hospital, Amausi, Lucknow, and is working in the capacity of an associate professor in the department of radiology. Hence, all measurements were made by the research scholar himself to reduce possible measuring errors made by different people. Along with the measurement at the time, simultaneously obtained values of maximum transverse diameter and minimum anteroposterior diameter at each particular vertebral level (T3 to T12) were entered in a data entry excel sheet for each patient. The same methodology was adopted for both the normal (control) and COPD (case study) groups. Separate master data entry sheets were created for the study's case and control groups. All imaging exposures were performed while patients were instructed to take inspiration and hold their breath for the duration.

Maximum transverse internal diameter and minimum anteroposterior internal diameter in centimeters were measured with a precision of.01 cm using a PACS integrated with a CT machine for digital images in the axial plane. If the CT image was on a console and was of value for study, that was also considered. The digital Vernier caliper that was used to measure both diameters had a precision of.01 mm when used for measurement. Haller`s index for each particular measured vertebral level was calculated by dividing the maximum transverse internal diameter by the minimum anteroposterior internal diameter. As the Haller`s index is a ratio, it has no unit. The Haller index was calculated by the following formula: -

$$\text{Haller index (HI)} = \frac{\text{Width (Internal maximum transverse diameter) in CM (X)}}{\text{Height (Internal minimum Antero-posterior diameter) in CM (Y)}}$$

The same methodology and formula for the calculation of Haller`s index were adopted for both the case and control groups. For standardization and correctness, periodic training and consultation were regularly made with the study supervisor (guide) and co-supervisor (co-guide) by the research scholar.

Standard axial CT thorax images at each intended measured vertebral level from T3 to T12 for the normal anatomy of thoracic structures that were used to decide the specified vertebral level in adults are prepared under the supervision of a supervisor (guide) and co-supervisor (co-guide). The measurement and anatomical structures are always considered just above the lower border. The axial CT thorax image for reference and comparison -

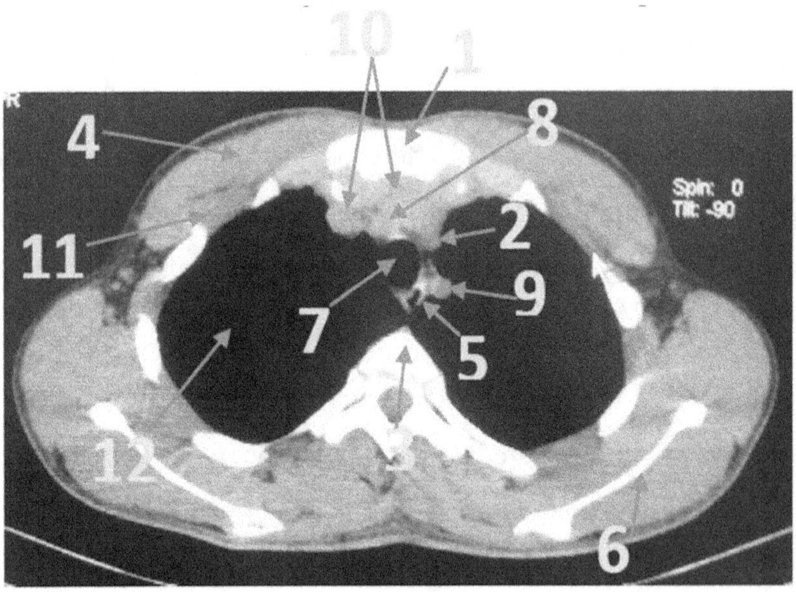

Figure 3.4 CT thorax image of a 20 years old male in axial plane at T3 showing normal anatomy

1. Manubrium
2. Left common carotid
3. T3 vertebra
4. Pectoralis major muscle
5. Esophagus
6. Scapula
7. Trachea
8. Brachiocephalic artery
9. Left subclavian artery
10. Brachiocephalic veins
11. Pectoralis minor muscle
12. Right upper lung

Note: At T3 level, all vessels are crowded with each other. The first rib's costal cartilage articulates with the manubrium. The esophagus is located just in front of and to the left of the vertebral body of T3.

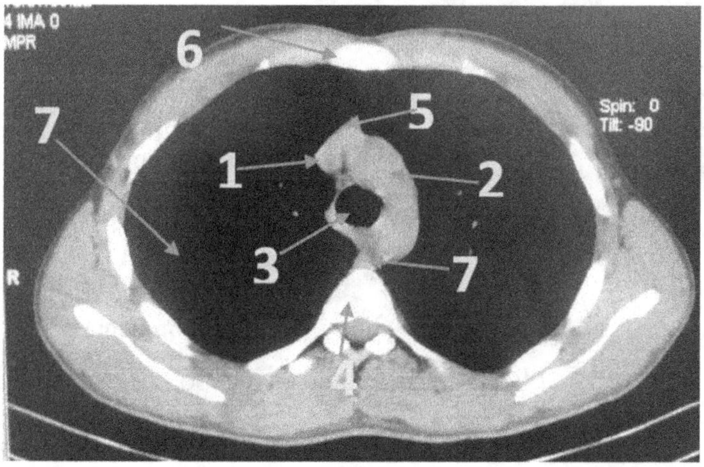

Figure 3.5 CT thorax image of a 20 years old male in axial plane at T4 vertebral level showing normal anatomy

1. Superior vena cava
2. Arch of aorta
3. Trachea
4. T3 vertebra
5. Remnant of thymus
6. Sternum
7. Right lung upper lobe

Note – at T4 level main characteristic structures are superior Vena cava, arch of aorta and remnant of thymus gland.

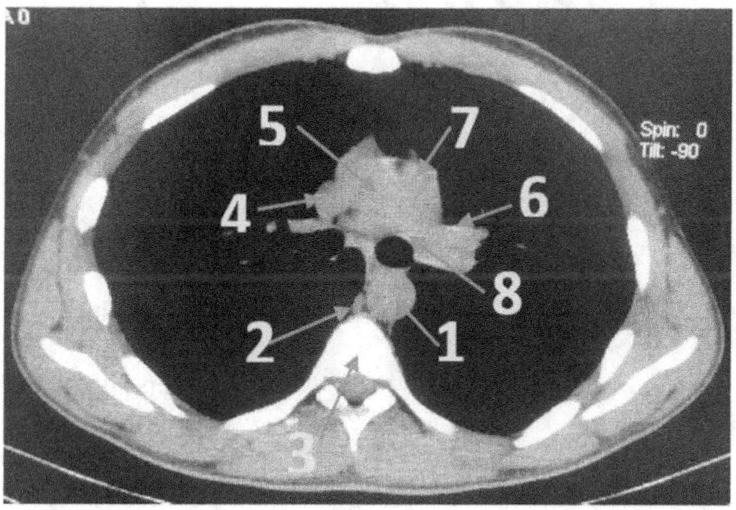

Figure 3.6 CT thorax image of a 20 years old male in axial plane at T5 vertebral level showing normal anatomy

1. Descending aorta
2. Esophagus
3. T5 vertebra
4. Superior vena cava
5. Ascending Aorta
6. Descending left recurrent laryngeal nerve
7. Left vagus
8. Ascending left recurrent

Note – Superior vena cava and ascending aorta in front of trachea, descending aorta and esophagus behind the trachea to the left just in front of body of T5. Ascending (8) recurrent laryngeal, descending laryngeal nerve (6), left vagus to the left of median axis.

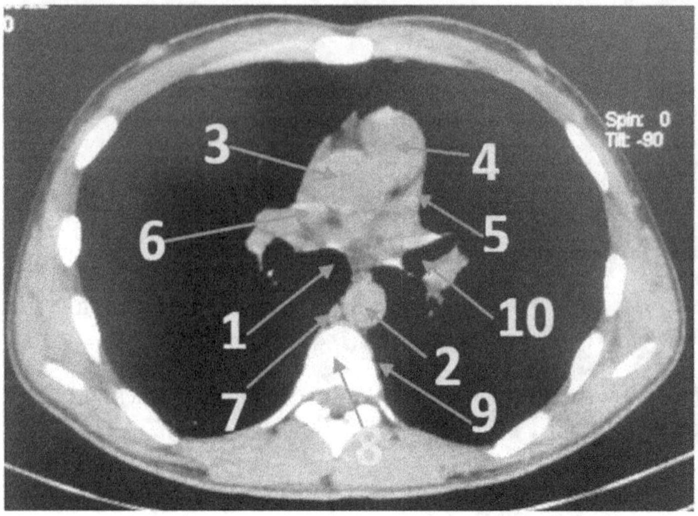

Figure 3.7 CT thorax image of a 20 years old male in axial plane at T6 vertebral level showing normal anatomy

1. Right upper lobe bronchus
2. Descending aorta
3. Ascending aorta
4. Pulmonary artery
5. Left phrenic nerve
6. Superior vena cava
7. Oesophagus
8. T6 vertebra
9. Sympathetic chain
10. Left bronchus

Note-At T6 level Pulmonary artery (4) lies anteriorly and to The left of ascending orta in the median plane. Right upper Lobe bronchus behind superior vena cava to the right side. Left Phrenic nerve between pulmonary artery and Left Bronchus to the left side of median plane.

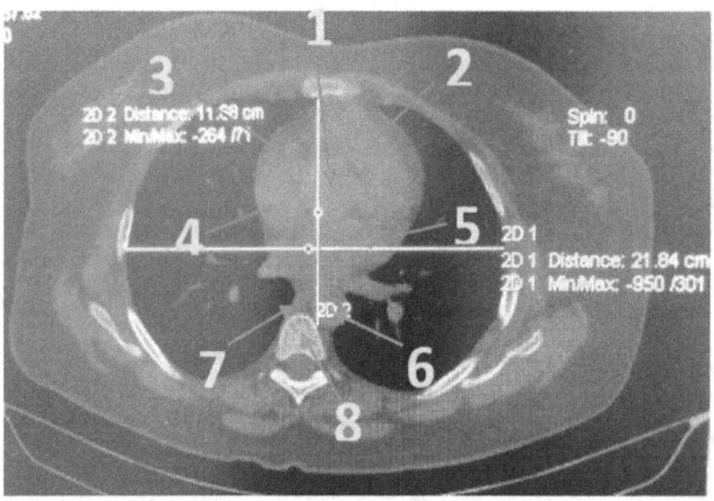

Figure 3.8 CT thorax image of a 20 years old male in axial plane at T7 vertebral level showing normal anatomy

1. Ascending aorta
2. Right ventricle
3. Right atrial appendage
4. Right atrium
5. Left atrium
6. Descending Aorta
7. Oesophagus
8. T7 vertebra

Note- At T7 level both right atrium and right auricle seen to the right of ascending aorta, ascending aorta is in center and right ventricle above and to the left lateral of the ascending aorta. Left atrium to the opposite of right atrium at same level.

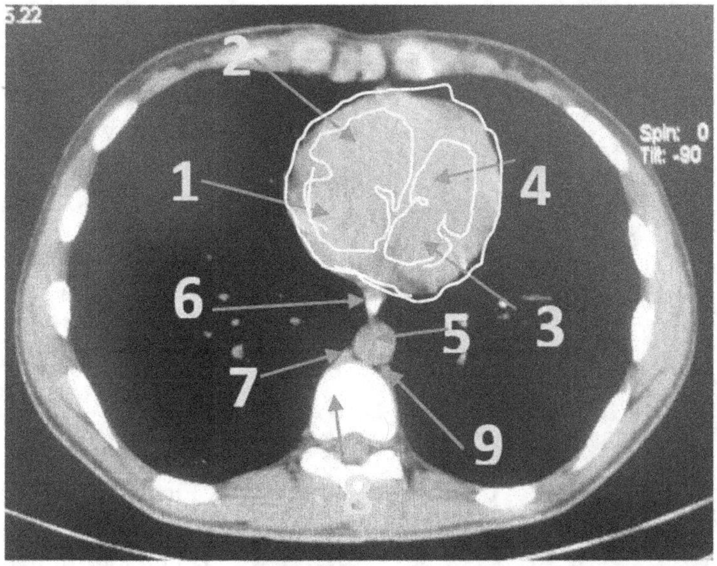

Figure 3.9 CT thorax image of a 20 years old male in axial plane at T8 vertebral level showing normal anatomy

1. Right atrium
2. Right ventricle
3. Left atrium
4. Left ventricle
5. Descending aorta
6. Oesophagus
7. Azygos vein
8. T8 vertebra
9.Sympathetic chain

Note- At T8 level all four chambers of the heart are discernible, behind left atrium esophagus is present, more posterior and to the left is descending aorta, on the right side and in front of body of T9 azygous azygos vein and to the left sympathetic chain. Also posteriorly base of the heart is related to T5- to T8 in supine position as CT is taken in supine position.

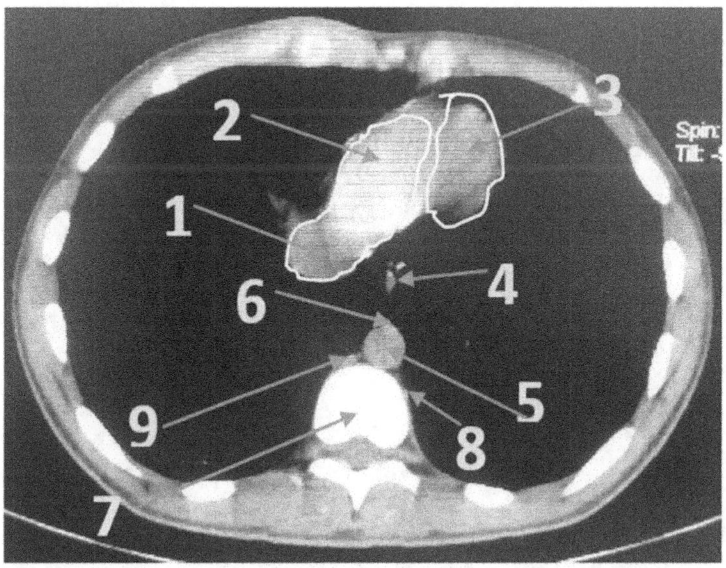

Figure 3.10 CT thorax image of a 20 years old male in axial plane at T9 vertebral level showing normal anatomy

1. Right atrium
2. Right ventricle
3. Left Ventricle
4. Left atrium small part
5. Descending aorta
6. Oesophagus
7. T9 vertebra
8. Sympathetic chain
9. Greater splanchnic

Note- At T9 level both right and left ventricles are seen Right atrium also seen but only small part of left atrium Oesophagus just anterior to the descending aorta, greater Splanchnic nerve to right of descending aorta.

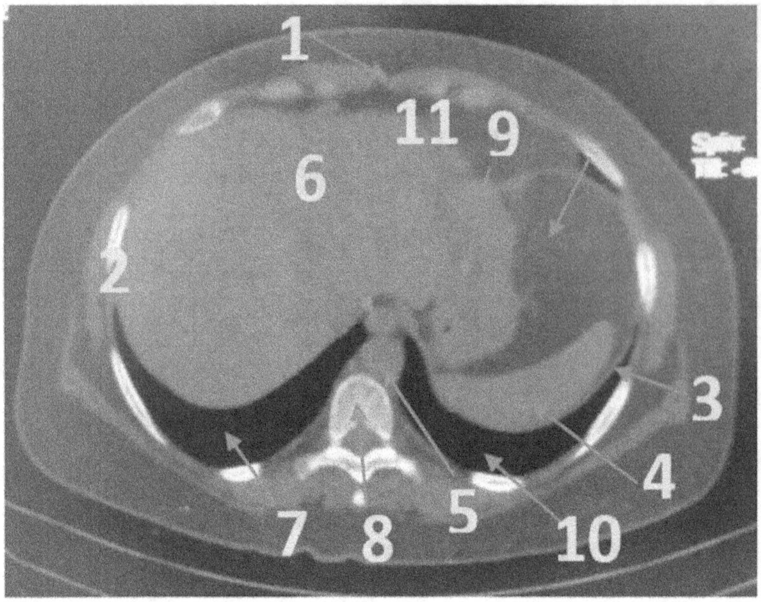

Figure 3.11 CT thorax image of a 38 years old female in axial plane at T10 vertebral level showing normal anatomy

1. Linea alba
2. Right lobe of liver
3. Left diaphragm
4. Left spleen
5. Descending thoracic aorta
6. Inferior vena cava
7. Right lower lobe of lung
8. T10 vertebra
9. Findus of Stomach
10. Left lower lung
11. Left lobe of liver

Note- At T10 anteriorly between rectus abdomins muscle linea alba is present. on the right side right lobe of liver occupies most of the part and small part of right lower lobe of lung. in front of body of T10 vertebra and to the left descending thoracic aorta and in front and slight to the right of aorta and between liver inferior vena cava is discernible. to the left of median plane stomach and spleen is seen.

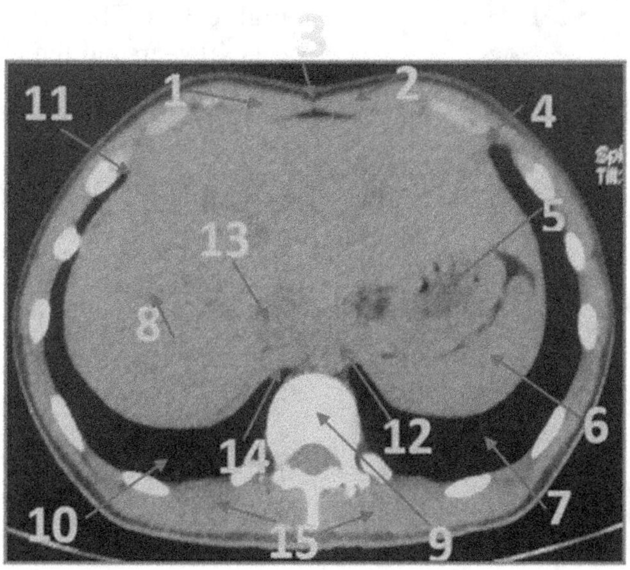

Figure 3.12 CT thorax image of a 20 years old male in axial plane at T11 vertebral level showing normal anatomy

1. Right rectus abdomins muscle
2. Left rectus abdomins muscle
3. Linea alba
4. Left diaphragm
5. Stomach
6. Spleen
7. Left lower lobe of lung
8. Right lobe of liver
9. T11 vertebra
10. Right lower of lung
11. Right diaphragm
12. Descending thoracic aorta
13. Inferior vena cava
14. Right crus of diaphragm
15. Errector spinae muscle

Note- At T11 level in front of body of T11 descending aorta (12) To the left and inferior vena cava (13) is discernible in right lobe of liver to the right of median plane. Spleen and stomach is close to each other on left side. Spleen appears in maximum length in its long axis.

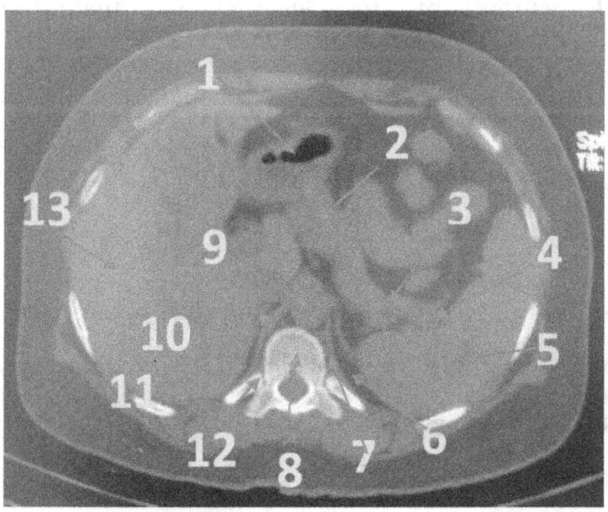

Figure 3.13 CT thorax image of a 38 years old female in axial plane at T12 vertebral level showing normal anatomy

1. Stomach
2. Upper part of Pancreas
3. Splenic Vessel
4. Hilum of spleen
5. Spleen
6. Left pleural cavity
7. Left crus of diaphragm
8. T12 vertebra
9. Descending thoracic aorta
10. Right crus of diaphragm
11. Right pleural cavity
12. Twelfth costovertebral joint
13. Liver

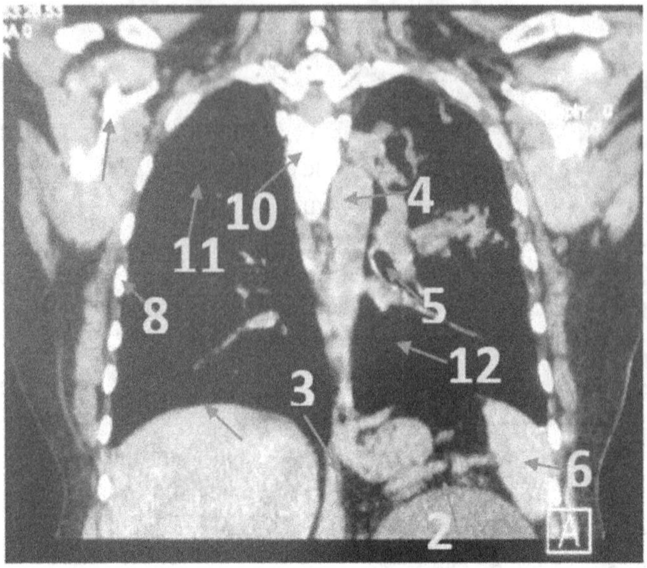

Figure 3.14 CT thorax image of a 40 years old female in coronal plane showing normal anatomy

1. Right dome of diaphragm
2. Left dome of diaphragm
3. Oesophagus
4. Arch of aorta
5. Left bronchus
6. Spleen
7. Clavicle
8. Ribs
9. Scapula
10. Thoracic vertebrae
11. Right lung
12. Left lung

In current study, the case group (COPD) consisted of chest computed tomography images of patients with chronic obstructive pulmonary disease (COPD), Chronic bronchitis, emphysema, bronchial asthma, and bronchiectasis were diseases included in the COPD group, and the COPD patient was separated from other diseases based on his HRCT findings. As these diseases progress, the alveoli become obstructed and air-trapped, destroying the alveolar walls. In extreme circumstances, these diseases also cause chest deformation in pigeons. Despite the fact that more studies are using spirometry and pulmonary function testing (PFT) to identify and classify patients. We used HRCT findings showing anatomically recognized lesions and irreversible changes in the lung parenchyma that could lead to changes in chest dimensions.

The discovery of high-resolution computed tomography has made it possible to recognize chronic obstructive pulmonary disease more reliably and diagnose more accurately, even in its milder forms [163]. Previous studies of pulmonary function testing in patients with asthma have reported an accelerated rate of decline in forced expiratory volume in one second (FEV1) in some groups[164, 165]. The cause of decline in FEV1 was mechanical changes and remodeling, that may contribute to[166, 167, 168] small respiratory airways of the lungs. Spirometry testing is insensitive to detecting small (<2 mm) airway diseases. This is because these small airways account for only about one-sixth of the total airflow resistance [169] in COPD. All patients in the study group were classified as a part of the COPD group if the HRCT findings suggestive of chronic obstructive pulmonary disease (COPD). The radiologist's final report was the sole and final source of documentation and categorization as COPD group. The investigator did not make his own observations to make a diagnosis. The majority of investigators agree that HRCT is accurate and sensitive for detecting all obstructive pulmonary disease. Even the earliest lesions of emphysema can be accurately diagnosed without any doubt. Even the earliest lesions of emphysema can be accurately diagnosed without any doubt [169]. The standard diagnostic criteria used by radiologists are.

1. HIGH-RESOLUTION COMPUTED TOMOGRAPHY CRITERIA FOR DIAGNOSIS OF EMPHYSEMA

(i) High-resolution computed tomography scans are highly sensitive in determining whether emphysema is present or not and its severity too[170, 171, 172, 173].

(ii) If the areas of non-peripheral and peripheral mild attenuation belonged to the subjective description of emphysema symptoms on computed tomography (CT) and high-resolution computed tomography (HRCT)[170].

(iii) If HRCT findings were suggestive of oblivious emphysematous bullae[171].

(iv) If pulmonary vascular shaving off (pruning) and distortion were [present170, 172].

(v) If low-attenuation non-peripheral and tangential regions, as well as cut back (pruning) and distortion of pulmonary vessels[170, 172] were among the interpretive features of computed tomography and high-resolution computed tomography emphysema images

(vi) If high resolution computed tomography (HRCT) suggestive of vascular disruption[174].

(vii) Centrilobular emphysema- presence of localize area of abnormally low attenuation without surrounding walls or if alveolar wall diameter was less or equal to 1mm diameter, within a homogeneous background of lung parenchyma, and not associated with fibrosis[175].

(i) Pan lobular emphysema: On chest CT scans, this type of emphysema appears as a broad, uniformly weak area, generally distributed all over the lung fields, particularly in the lower lobes and smaller than normal pulmonary vessels [175].

(ii) Para septal emphysema: This condition manifests as numerous small subpleural air pockets a few millimeters to a centimeter in diameter[176].

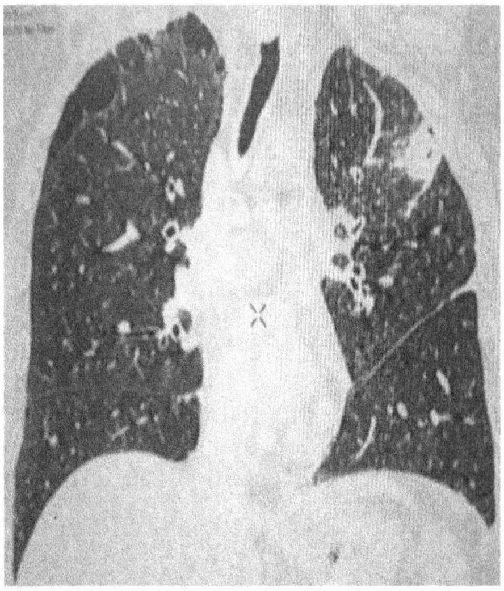

Figure no-3.15 Coronal view of lungs of a 45 years male showing Emphysema of Right Middle and upper peripheral lobe

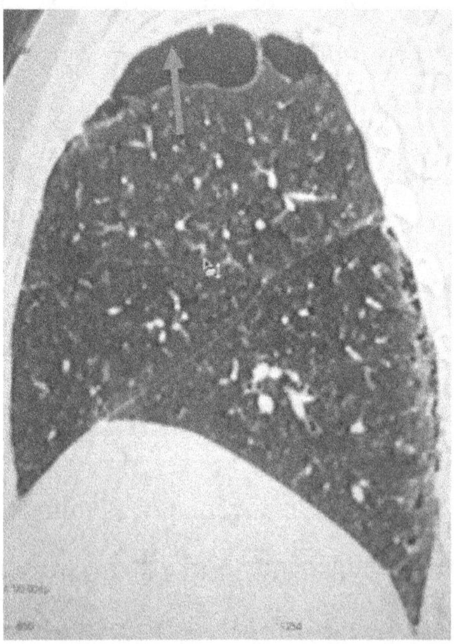

Figure no- 3.16

Sagittal view of lungs of a 45 years male showing Emphysema of Right Middle and upper peripheral lobe

2. HIGH RESOLUTION COMPUTED TOMOGRAPHY CRITERIA FOR DIAGNOSIS OF BRONCHITIS

When HRCT findings indicated wall thickness of small to medium-sized airways and nearby larger bronchi, there was an increase in measurements of more than 1.28 ± 0.11 mm [177, 178].

3. HIGH RESOLUTION COMPUTED TOMOGRAPHY CRITERIA FOR DIAGNOSIS OF BRONCHIECTASIS-

(i) If a circular sign was found on HRCT, indicating a cross-section of a dilated bronchus filled with air adjacent to a smaller opacity indicating pulmonary artery involvement[179], it is an early sign of bronchiectasis on CT.

(iii) Cylindrical, venous, and cystic bronchiectasis are the three main morphological forms of bronchiectasis that can be identified on a CT scan and many have a combination of these three classical forms[180].

(iv) On HRCT, CT findings related to bronchiectasis based on the following 3 criteria were observed: (1) absence of thin bronchial walls; (2) bronchiectasis when its internal diameter is at least 110% larger than the nearby pulmonary artery; or (3) imaging of a peripheral bronchus within 1 cm of the lateral pleural surface or the nearby mediastinal pleural surface[181].

(v) Cylindrical bronchiectasis was defined based on arterial wall thickness, and the ratio of bronchiectasis and wall thickening was found to be greater than 1 in HRCT[182].

(vi) If CT shows thin-walled cysts that may contain fluid, this is defined as cystic bronchiectasis and is detected in subsequent axial sections, either in a composite pattern or in a tree order [183].

(vii) To evaluate for bronchiectasis, six CT imaging criteria were also considered. For example: (1) bronchiectasis; (2) peri bronchial wall thickening; (3) the number of bronchiectasis segments; (4) the number of bubbles; (5) the emphysema segment number; and (6) the criteria for associated small airway disease[184].

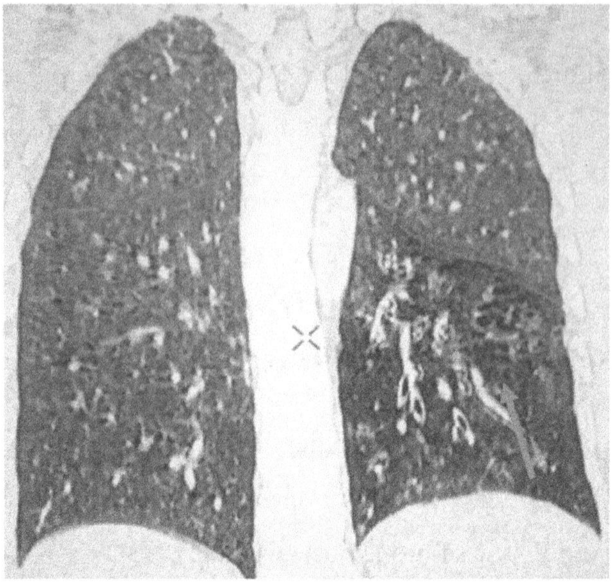

Figure no-3.17

Coronal view of lungs of a 27-year male bronchiectasis of left lower lobe

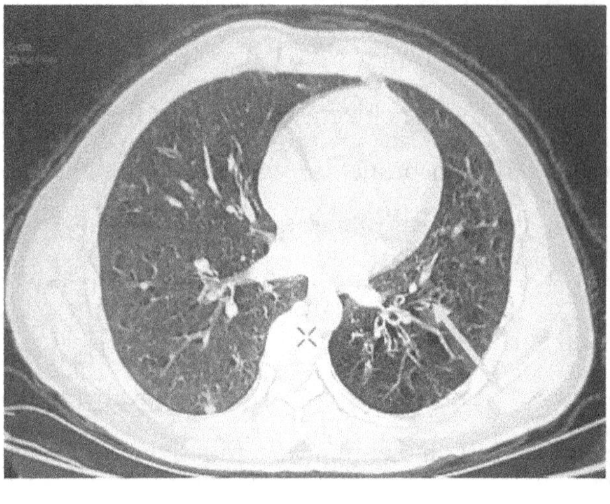

Figure no-3.18

CT Thorax image in axial plane of a 27-year male showing bronchiectasis of the left lower lobe

4. HIGH RESOLUTION COMPUTED TOMOGRAPHY CRITERIA FOR DIAGNOSIS OF BRONCHIAL ASTHMA-

There is no doubt that HRCT scans are sensitive to emphysema changes in the lungs of smokers, but the specificity of this technique in asthmatic lungs is unknown. However, emphysema, or at least 'end-stage airspace enlargement', is a common finding on high-resolution computed tomography imaging of asthmatic lungs, and its severity may be proportional to the severity of asthma.

(i) Asthma is characterized by persistent inflammation of the airways, primarily affecting the medium and small bronchi[186].

(ii) (ii) bronchial thickening with edema, increased smooth muscle number, and increased mucous gland size[187]; these histological changes can be seen on high-resolution computed tomography as bronchial wall thickening and bronchial lumen narrowing[187].

(iii) (iii) High-resolution computed tomography revealed small opacities in the central lobules, air pockets, and areas of reduced density and vascularity [187].

(iv) High-resolution computed tomography scans taken at the end of inspiration revealed areas of reduced density and vascularity in approximately 20% of asthma patients[188, 189,190.]

High-resolution computed tomography in patients with asthma revealed mucus blockage or collapse. The severity of the reversible defect did not correlate with asthma severity. These data suggest that even people with mild asthma (scores 1-2 on the Aac scale of 19)[191] may have areas of mucous blockage that are virtually unnoticeable. Although these studies demonstrate the sensitivity of HRCT to detect mucus obstruction, the clinical significance of this anomaly is unknown.

COMPUTED TOMOGRAPHY

Conventional X-ray machines use radiation to create images of the body's bones and soft tissues on special X-ray film. Structures in the radiation path appear excessively during the process. In contrast, computed tomography measures X-rays inside the tissue and creates a tomogram or slice image on the screen. The body is broken into smaller parts by an imaging system. Conceptually, this is similar to cutting through the stone, clearly showing the distribution of dark and light matter across its surface. A detailed, high-contrast view of tissues inside the body is provided by medical imaging. This has significant benefits for various clinical studies.

The body is stretched and superimposed in geographic depth, so it can be viewed as a 3D model on the monitor. Very high image resolution allows you to see even the smallest blood vessels, including in and around the heart. Or in your brain, visualizing hairline fractures, organ changes, looking for tumors, and examining the heart are all suitable for CT scans. In addition, it is used in emergency rooms to quickly identify internal injuries, organize surgeries, and monitor patient recovery. An average CT scan takes less than 10 minutes from preparation to final image and is completely painless.

PRESENT DAY COMPUTED TOMOGRAPHY

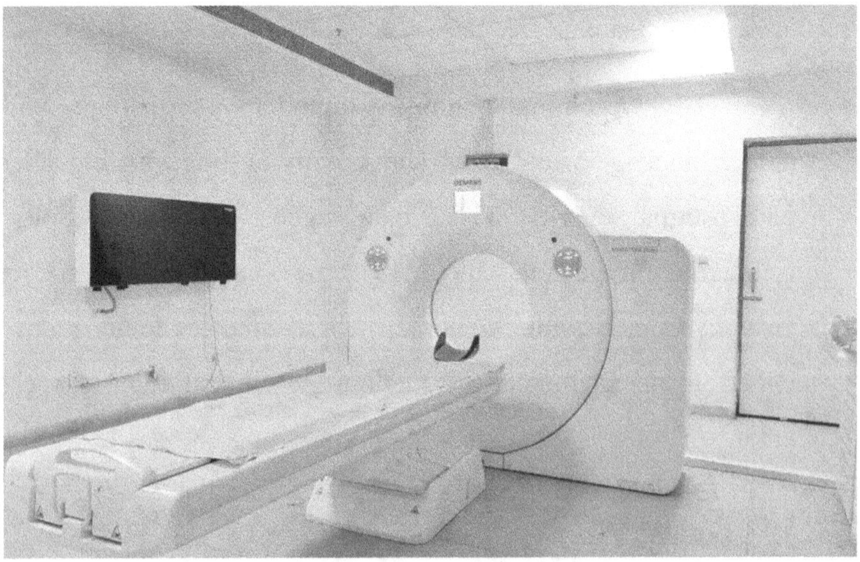

Figure no. 3.19

Siemens 64 slice computed tomography machine

Medical professionals refer to the CT scanner's ring-shaped entrance as the gate. Non-experts generally call them "doughnuts". The measuring equipment, consisting of an X-ray tube and detectors on both sides, is located inside the gate. A weighing machine, which usually weighs between 400 and 1600 kg, rotates around the subject in a few seconds. A fan-shaped x-ray beam penetrates the tube during this process and is weakened by harder tissue than soft tissue as it passes through the body. Light beams hit a "needle" on their way to the detector.

When X-rays are detected, the CT machine uses a very special combination of ceramics to convert the X-rays into light. The light is then converted into electricity by a photodiode, and a converter creates digital data from the analog signal and sends it to a computer for processing. Without noticeable delay, the computer converts the measurements into pictures of each area and a 3D model of the entire body. To examine organs such as the intestines, doctors can rotate the image of the body on the screen, zoom in and out, and even do a virtual "fly-through."

The optimal radiation dose for each scan is determined by analyzing the gross anatomy of each individual with modern CT equipment. Millisievert is used to measure radiation exposure (mSv). For people living in Germany, the average annual exposure to natural X-ray radiation is 2.4 millisieverts. Today's CT scanners can scan the lung with less than 0.1 mSv of radiation. A typical dose is 2-3 mSv.

DISCOVERY OF COMPUTED TOMOGRAPHY

In 1971, in England, a radiologist and technician once said, "I felt like a college athlete who had just hit a winning penalty." Two scientists owned the tomogram. This is a completely new type of X-ray image that shows the human heart in great detail. Radiologist James Ambrose, by studying the images and observing the patients' brains "with greater precision than we imagined," clearly identified the cerebral cortex, spaces filled with cerebrospinal fluid, and even white matter. I was able to do this. The new X-ray technology was practically invented by the technician Godfrey Hounsfield. Hunsfield was a pioneer of computed tomography (CT), ushering in the era of modern X-ray technology with his model "3D X-ray machine".

Imagine that certain rays from the sun reach the depth of a cone. While hiking with his friend Roger Walls in the mid-1960s, Godfrey Hansfield came up with a simple thought experiment that marked the beginning of computed tomography. If the detector was placed outside the pyramid, facing the sun, and moved during the day, the entire volume of the pyramid could be searched for hidden chambers. Containing their full contents, he knew that to get the third dimension, he had to scan the object through 180 degrees. Hansfield intended to superimpose this 3D data "like moving objects with a pencil sharpener" to make it easier to calculate the experimental measurements. Godfrey Hunsfield's theory was pure speculation until 1967. They were interesting puzzles that were the result of genuine scientific interest. Then his current project was abruptly terminated by his employer and his electronics company and label Electromagnetic Interface was shut down.

According to this legend, Godfrey Hunsfield had complete freedom in choosing his next project and was able to use the Beatles' success with electromagnetic interfaces to build the first CT scanner. In fact, it is the opposite. Hansfield had no original plans to enter the healthcare electromagnetic interface industry, so he had to scramble to fund his research. It took me months to convince my bosses, one of whom called the idea "crazy," but I finally won their trust. Hunsfield received about a quarter of the amount he requested to build the prototype. William E. Ingham, director of research at Interface Electromagnetics at the time, said the budget "did too much for too little".

Godfrey Hansfield and his small team of engineers did everything they could to save money. For example, they built the basic structure of a lab prototype from an old lathe and borrowed paper tape punches to store data. Each scan was recorded as 28,800 holes spaced more than 60 meters apart. Steven Bates, the team's software engineer, recalls that he had "never before, or since" worked on a project that used so many shortcuts and seemingly inadequate tools. Despite all the challenges, the team was able to turn Hounsfield's concept into a functional scanner in just one year. They did the first CT scan in early 1969 using a 9-day exposure time and images of some plastic lab equipment. Thanks to EMI funding and support from the UK Department of Health and Social Security, they used better components and reduced the exposure time to around nine hours within a few months. A few months ago they cut most of it down to about 5 minutes. It was the first major clinical prototype at the James Ambrose Ward at the Atkinson Morley Hospital, Wimbledon, on 1 October 1971, and was already shared with current CT scanners. The first patient's brain scan data was recorded on magnetic tape and transported by car to an electromagnetic interface laboratory about 20 kilometers away. This was because there was no computer to create the images in the system and throughout the hospital. Two days later, Hansfield and Ambrose were jumping up and down with joy, holding the diagnostic image. They began further research and presented their findings at the Congress of Radiology in London on April 20, 1972. This topic has sparked the greatest interest in medical X-ray technology in the last few decades.

William E. Ingam later remembered. Computed tomography essentially changed the course of medical x-ray technology development overnight. Thanks to Godfrey Hansfield's invention, pneumoencephalography, which had been common until then, was completely unnecessary. Cerebrospinal fluid is removed from the lumbar region of the patient's spine and replaced with air during this often painful and difficult procedure to allow x-rays of the brain. Side effects such as vomiting, seizures and even meningitis with pneumoencephalography were frequent and included hospitalization for several days. Conversely, patients can be scanned with CT scans without any discomfort, and even the first prototype of the electromagnetic interface produced much more precise and contrast-rich results. In addition to being featured in medical journals, almost every British publication and newspaper featured Godfrey Hunsfield's findings and new radiographs. Coverage of these issues began in The Times, followed by The Guardian, The Daily Telegraph and the British Broadcasting Corporation.

Another 17 companies have begun work on developing CT scanners in addition to electromagnetic interfaces. Since its inception, computed tomography has undergone rapid technological development. Amazingly, Godfrey Hunsfield accurately predicted many of the innovations that would appear over the next four decades, such as the introduction of the multi-slice CT scanner in 1998. Anticipating some of the key innovations that would shape computed tomography over the years, Siemens Health engineers it was impossible.

The trip to England was the beginning of computed tomography for Siemens Health engineers. In the summer of 1972, Oskar Doenisch, head of innovation in Siemens' medical technology division, and Friedrich Göden, head of X-ray development at Siemens, visited the Electromagnetic Interface Research Center to discuss the company's advances in computing technology. The visit was "very informative," according to Godden's recollections of the planned CT scan. "The food was great and the pioneer of computed tomography, Godfrey Hansfield, was in attendance. He had a huge impact on me. His explanations were fascinating, including that it took nine days to collect the measurements for each image in the first place.

RESULTS

Table – 1: Test of Normality

Parameter	Kolmogorov-Smirnov Test of Normality		
	Statistic	df	p-value
T3_Maximum Internal Transverse Diameter	.056	384	.200*
T4_Maximum Internal Transverse Diameter	.031	384	.200*
T5_Maximum Internal Transverse Diameter	.027	384	.200*
T6_Maximum Internal Transverse Diameter	.032	384	.200*
T7_Maximum Internal Transverse Diameter	.040	384	.172
T8_Maximum Internal Transverse Diameter	.034	384	.200*
T9_Maximum Internal Transverse Diameter	.034	384	.200*
T10_Maximum Internal Transverse Diameter	.030	384	.200*
T11_Maximum Internal Transverse Diameter	.059	384	.099
T12_Maximum Internal Transverse Diameter	.029	384	.200*
MaximumInternal Transverse Diameter	.023	384	.200*
T3_Minimum Anteroposterior Diameter	.055	384	.200*

T4_Minimum Anteroposterior Diameter	.043	384	.200*
T5_Minimum Anteroposterior Diameter	.043	384	.092
T6_Minimum Anteroposterior Diameter	.044	384	.069
T7_Minimum Anteroposterior Diameter	.042	384	.108
T8_Minimum Anteroposterior Diameter	.054	384	.200*
T9_Minimum Anteroposterior Diameter	.065	384	.045
T10_Minimum Anteroposterior Diameter	.064	384	.056
T11_Minimum Anteroposterior Diameter	.036	384	.200*
T12_Minimum Anteroposterior Diameter	.054	384	.200*
Minimum Anteroposterior Diameter	.062	384	.073
T3_Haller Index	.024	384	.180
T4_Haller Index	.061	384	.076
T5_Haller Index	.071	384	.020
T6_Haller Index	.079	384	.055
T7_Haller Index	.075	384	.060
T8_Haller Index	.057	384	.200*
T9_Haller Index	.036	384	.200
T10_Haller Index	.042	384	.200*
T11_Haller Index	.050	384	.200*
T12_Haller Index	.066	384	.053
HallerIndex	.075	384	.061

The table No.1 shows the results of the Kolmogorov-Smirnov test of normality performed on various parameters of interest. This test is used to determine if a given dataset follows a normal distribution or not.

The table reports three values for each parameter: the statistic, the degrees of freedom (df), and the p-value. The statistic represents the maximum absolute difference between the empirical cumulative distribution function (ECDF) of the data and the theoretical cumulative distribution function (CDF) of a normal distribution. The df is the degrees of freedom for the test, which is based on the number of observations in the dataset. The p-value is the probability of obtaining a test statistic as extreme as the one observed, assuming that the null hypothesis (i.e., the data follows a normal distribution) is true.

If the p-value is less than the significance level (often set at 0.05), then we reject the null hypothesis and conclude that the data does not follow a normal distribution. On the other hand, if the p-value is greater than the significance level, we fail to reject the null hypothesis, and we conclude that the data is normally distributed.

In this case, for most parameters, the p-values are greater than 0.05, indicating that we fail to reject the null hypothesis and conclude that the data is normally distributed. However, for a few parameters, such as T5_Haller Index, T9_Minimum Anteroposterior Diameter, and T12_Haller Index, the p-values are less than 0.05, suggesting that we should reject the null hypothesis and conclude that the data does not follow a normal distribution.

❖ Age Distribution of control and Cases

Table – 2: Age Distribution among Case and control groups

Age	Normal group		COPD group		Total	
	No.	%	No.	%	No.	%
18 - 29 yr	46	24.0%	19	9.9%	65	16.9%
30 - 39 yr	39	20.3%	22	11.5%	61	15.9%
40 - 49 yr	35	18.2%	33	17.2%	68	17.7%
50 - 59 yr	39	20.3%	50	26.0%	89	23.2%
60 - 70 yr	33	17.2%	68	35.4%	101	26.3%
Mean±SD	42.34±14.77 yr		51.08±13.46 yr		46.71±14.77 yr	

Table No. 2 Demonstrate the result-based age distribution of the control group and case group illustrated that the mean age of the normal group was 42.34±14.77 yr, the mean age of the COPD group was 51.08±13.46 yr, and overall, the mean age of the study cases was 46.71±14.77 yr. Furthermore, in the normal group, the most subjects were found in the age group 18-29 years, with 46 participants (24%), followed by 30–39 years, with 39 participants (20.3%), 40–49 years, with 35 participants (18.2%), 50–69 years, with 39 participants (20.3%), and 60–70 years, with 33 participants (17.2%), while in the COPD group, the most subjects were found in the age group 60–70 years, with 68 participants (35.4%pants (20.3%), 40–49 years, with 35 participants (18.2%), 50–69 years, with 39 participants (20.3%), and 60–70 years, with 33 participants (17.2%), while in the COPD group, the most subjects were found in the age group 60–70 years, with 68 participants and constituted 35.4%, followed by the 50-59 years age group with 50 participants (26.0%), the 40-49 years age group with 33 participants (17.2%), the 30-39 years age group with 22 participants (11.5%), and the 18-29 years age group with 19 participants (9.9%) (refer to Table-2)

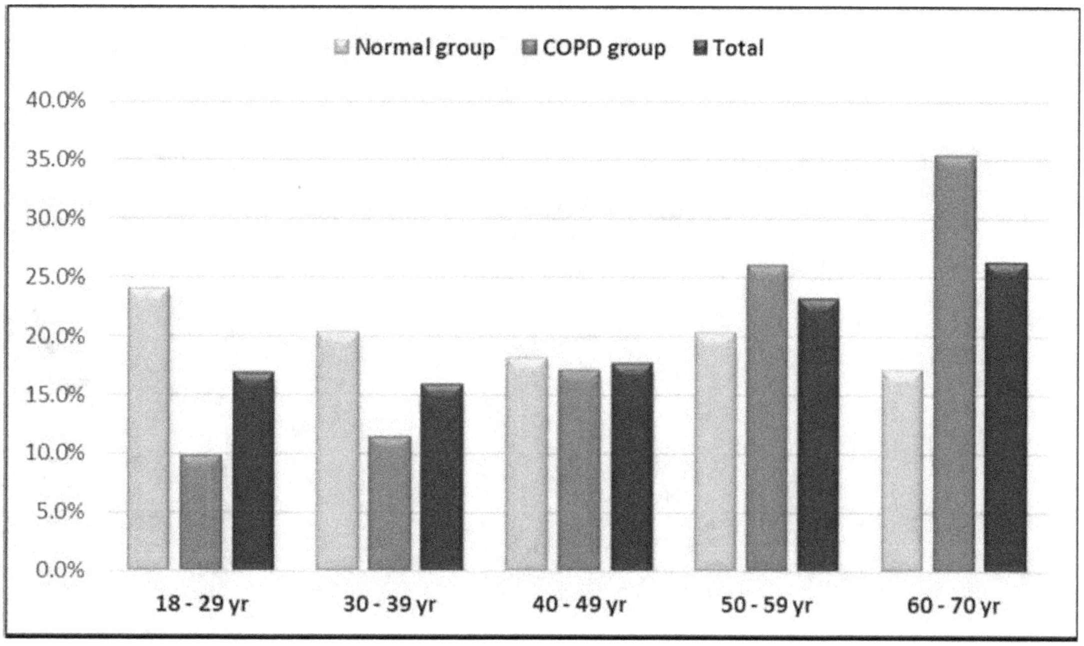

Graph No. 1.01: Bar graph showing the age distribution of the control and case groups

❖ Sex Distribution of Case and control groups

The result-based Sex Distribution of Control and Cases illustrated that in the normal group and COPD groups, participants had equal numbers of male and female subjects (Table 3).

Table – 3: Sex Distribution of Case and control groups

Sex	Normal group		COPD group		Total	
	No.	%	No.	%	No.	%
Male	96	50.0%	96	50.0%	192	50.0%
Female	96	50.0%	96	50.0%	192	50.0%
Total	192	100.0%	192	100.0%	384	100.0%

Table No. 3 demonstrated that in the control and COPD groups, participants were equal in number. Males and females were also equal in number and percentage.

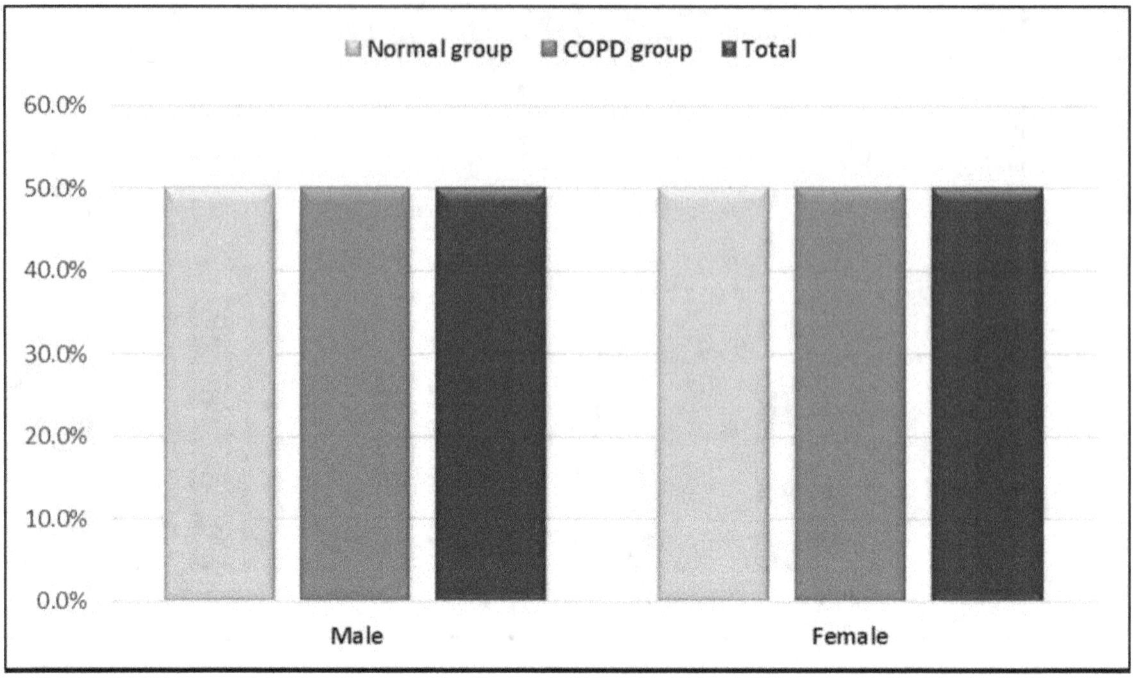

Graph No. - 1.02: Bar graph showing distribution of sex among Control and case groups.

- ❖ **Descriptive Summary and Comparison of Maximum Internal Transverse Diameter between Normal & COPD groups among Males (Table- 4)**

The results-based summary and comparison of maximum internal transverse diameter among males in control and case groups using an unpaired T test illustrated that a significant association was found in both groups with T4 ($t = 2.03$; $p = 0.044$), T5 ($t = 2.22$; $p = 0.028$), and T6 ($t = 2.10$; $p = 0.037$) (Table 4).

Table 4: Descriptive Summary and Comparison of Mean Maximum Internal Transverse Diameter Per-Vertebral Levels between Normal and COPD Groups among Males

Male - Maximum Internal Transverse Diameter	Normal group		COPD group		unpaired t test	
	Mean	SD	Mean	SD	t-value	p-value
T3	19.23	1.93	18.69	1.89	1.96	0.052
T4	21.32	1.72	20.85	1.47	2.03	**0.044**
T5	22.51	1.61	22.05	1.23	2.22	**0.028**
T6	23.25	1.56	22.82	1.25	2.10	**0.037**
T7	23.71	1.55	23.31	1.31	1.91	0.058
T8	24.03	1.63	23.70	1.89	1.29	0.200
T9	24.61	1.77	24.64	1.66	-0.14	0.886
T10	25.21	1.78	25.50	1.82	-1.11	0.267
T11	25.79	1.74	26.16	1.85	-1.43	0.155
T12	25.92	1.82	26.39	2.20	-1.62	0.107

Table no. 4 demonstrated Transverse diameters were shortest at cranial level T3 and greatest at caudal level T12 in both the control and COPD groups. Among both groups, successive increments were noted from T3 to T12. The vertebral levels T3 to T8 transverse diameters were slightly higher in the control group as compared to the COPD group. While from T9 to T12, transverse diameters were slightly higher in the COPD group. The difference was statistically significant in transverse diameter between normal and COPD males at vertebral levels T4, T5, and T6. Normal male had noticeable higher values at T4, T5, and T6 as compared to COPD.

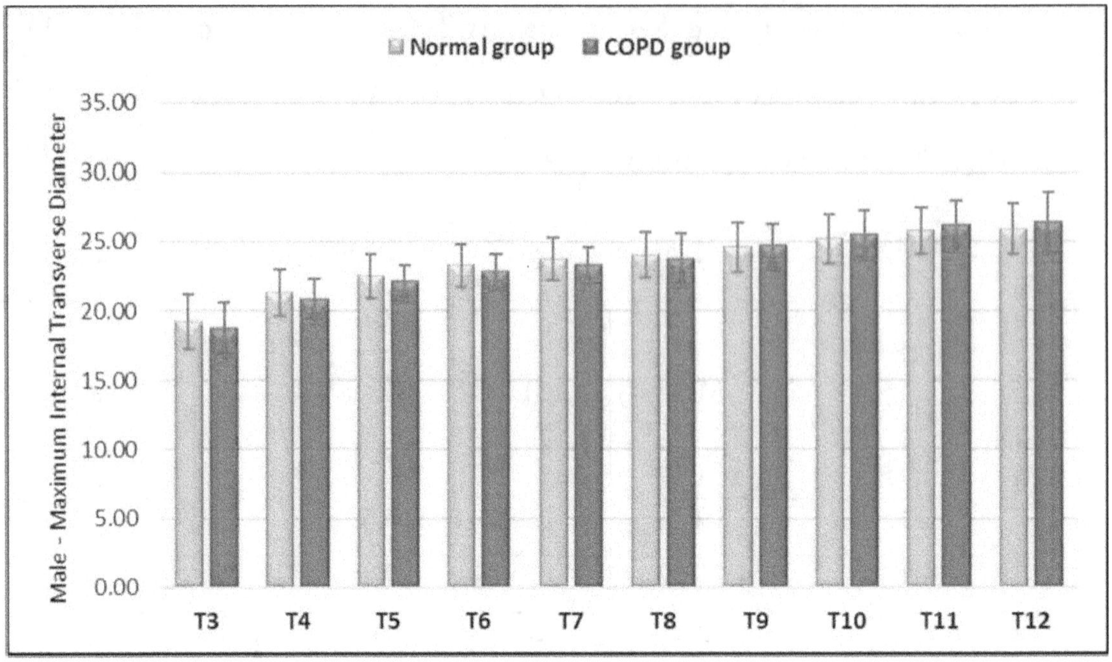

Graph No. 1.03: A bar graph comparing maximum transverse diameters at various per vertebral levels.

- **Descriptive Summary and Comparison of Minimum Anteroposterior Diameter between Normal and COPD groups among Males (Table-5)**

The results-based summary and comparison of minimum anteroposterior diameter among male controls and cases using an unpaired T test illustrated that a significant association was found in both groups with T3(t=-9.02; p<0.001), T4 (t=-10.26; p<0.001), T5 (t=-13.23; p<0.001), T6(t=-14.52; p<0.001), T7(t==15.26; p<0.001), T8(t=-14.68; p<0.001), T9(t=-13.57; p<0.001), T10(t=-12.66; p<0.001), T11(t=-11.25; p<0.001) and T12(t=-8.93; p<0.001) (Table 5).

Table 5: Descriptive Summary and Comparison of Minimum Anteroposterior Diameter between Normal and COPD Groups among Males

Male - Minimum Anteroposterior Diameter	Normal group		COPD group		unpaired t test	
	Mean	SD	Mean	SD	t-value	p-value
T3	5.57	0.89	7.02	1.30	-9.02	**<0.001**
T4	7.24	0.98	9.07	1.43	-10.36	**<0.001**
T5	8.48	0.99	10.71	1.31	-13.23	**<0.001**
T6	9.49	1.05	11.92	1.26	-14.51	**<0.001**
T7	10.18	1.12	12.77	1.23	-15.26	**<0.001**
T8	10.56	1.19	13.28	1.37	-14.68	**<0.001**
T9	10.71	1.32	13.74	1.74	-13.57	**<0.001**
T10	11.13	1.54	14.36	1.97	-12.66	**<0.001**
T11	11.58	1.71	14.67	2.08	-11.25	**<0.001**
T12	11.34	1.98	14.22	2.45	-8.93	**<0.001**

This table No. 5 demonstrated that the minimum anteroposterior diameters in the normal and COPD groups among males were shortest at cranial level T3 and highest at T11. Successive rising patterns were noted from T3 to T11, and a slight downfall was noted at T12 in both groups. An important fact was found: the minimum anteroposterior diameters were significantly higher in the COPD group as compared to the normal group at each vertebral level.

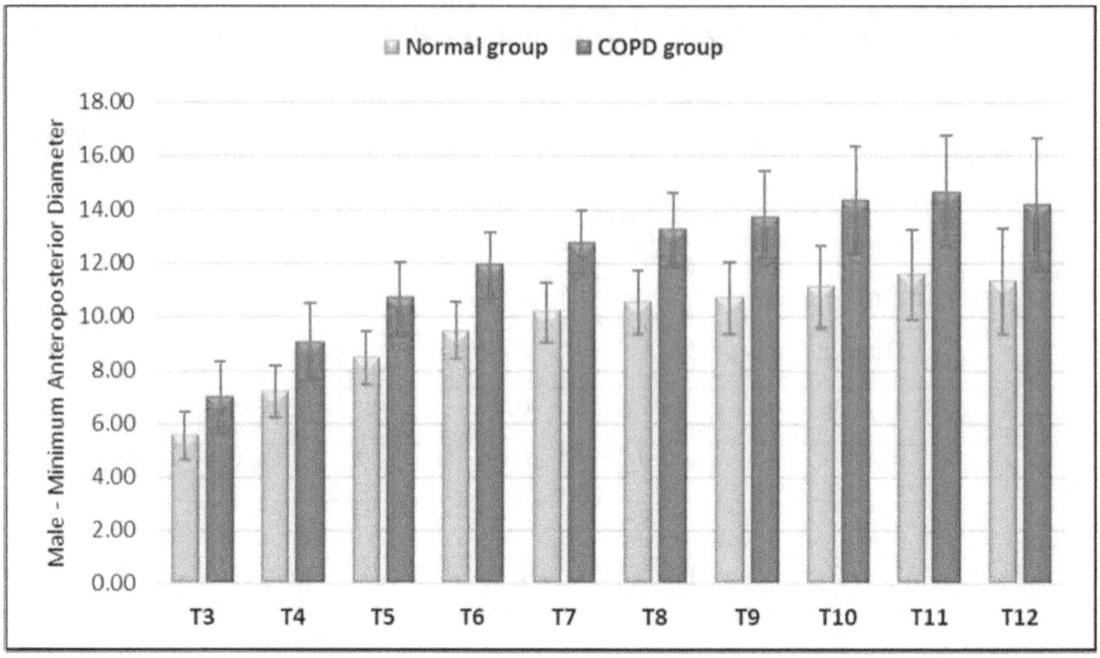

Graph No. 1.04: A bar graph comparing the minimum anteroposterior diameters at the vertebral level in males with normal and COPD.

❖ Descriptive Summary and Comparison of Minimum Anteroposterior Diameter between Normal & COPD groups among Males (Table-6)

The result-based summary and comparison of Haller indices among male controls and cases using an unpaired T test illustrated that a significant association was found in both groups with T3 (t=11.88; p<0.001), T4 (t=13.46; p<0.001), T5 (t=15.43; p<0.001), T6(t=16.67; p<0.001), T7(t=16.51; p<0.001), T8(t=14.98; p<0.001), T9(t=14.61; p <0.001), T10(t=13.69; p<0.001), T11(t=12.35; p<0.001) and T12(t=9.77; p<0.001) (Table 6)

Table 6: Descriptive Summary and Comparison of Haller Indices Between Normal and COPD Groups Among Males

Male - Haller Index	Normal group		COPD group		unpaired t test	
	Mean	SD	Mean	SD	t-value	p-value
T3	3.52	0.53	2.72	0.39	11.88	**<0.001**
T4	2.98	0.36	2.34	0.30	13.46	**<0.001**
T5	2.68	0.30	2.08	0.23	15.43	**<0.001**
T6	2.47	0.26	1.93	0.18	16.67	**<0.001**
T7	2.35	0.26	1.84	0.16	16.51	**<0.001**
T8	2.30	0.27	1.80	0.19	14.98	**<0.001**
T9	2.32	0.28	1.81	0.19	14.61	**<0.001**
T10	2.30	0.30	1.80	0.19	13.69	**<0.001**
T11	2.27	0.31	1.81	0.19	12.35	**<0.001**
T12	2.34	0.34	1.90	0.28	9.77	**<0.001**

Table No. 6 shows that the Haller indices in the normal and COPD groups were highest at cranial vertebral level T3. In normal males, Haller indices were in a decreasing pattern from T3 to T11 except for T9, and again, T12 had a higher value than T11. In the COPD group, Haller indices were found in a decreasing pattern from vertebral level T3 to T8. Vertebral levels T9 and T11 had the same values, and T12 had a higher value as compared to T11. It was clear that the COPD group's Haller indices were always lower than the normal group's per vertebral level.

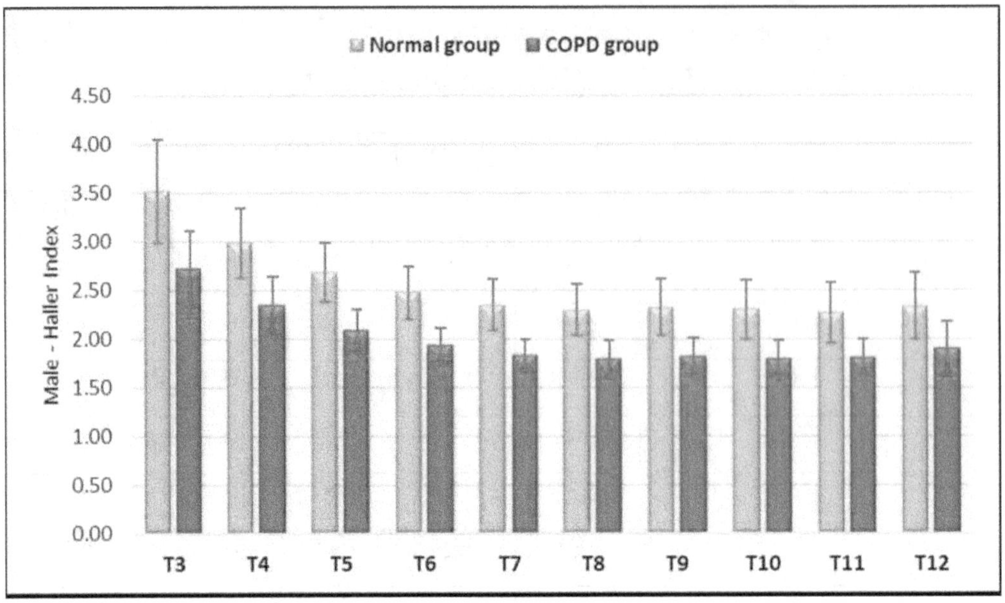

Graph No. 1.05: Bar graph representing the Haller index per vertebral level in the normal and COPD groups among males

- ### Descriptive Summary and Comparison of Maximum Internal Transverse Diameter between Normal & COPD groups among Females (Table-7)

The result-based summary and comparison of maximum internal transverse diameter between the control and COPD groups among females using an unpaired T test illustrated that a significant association was found in both groups with T3 (t=2.68; p=0.008), T4 (t=3.17; p=0.002), T5 (t=3.48; p=0.001), T6(t=3.69; p<0.001), T7(t=3.43; p=0.001) and T8(t=2.33; p=0.021) (Table 7).

Table 7: Descriptive Summary and Comparison of Maximum Internal Transverse Diameter between Normal and COPD Groups among Females

Female - Maximum Internal Transverse Diameter	Normal group		COPD group		unpaired t test	
	Mean	SD	Mean	SD	t-value	p-value
T3	17.25	1.88	16.46	2.17	2.68	**0.008**
T4	19.38	1.43	18.64	1.77	3.17	**0.002**
T5	20.49	1.26	19.79	1.50	3.48	**0.001**
T6	21.16	1.21	20.47	1.39	3.69	**<0.001**
T7	21.56	1.20	20.91	1.40	3.43	**0.001**
T8	21.82	1.22	21.36	1.50	2.33	**0.021**
T9	22.17	1.35	21.97	1.63	0.93	0.352
T10	22.70	1.49	22.75	1.76	-0.22	0.823
T11	23.05	1.66	23.35	1.75	-1.22	0.225
T12	23.16	1.77	23.64	2.16	-1.68	0.094

This table no. 7 demonstrated the shortest transverse diameter at cranial level T3 and the largest at T12. Also, an increasing pattern was seen at successive vertebral levels from T3 to T12 in the normal and COPD groups among females. It was also noted that COPD females had slightly lower values of transverse diameters at per vertebral levels from T3 to T9 as compared to normal females. while normal females had slightly lower values at T10 to T12 as compared to COPD females.

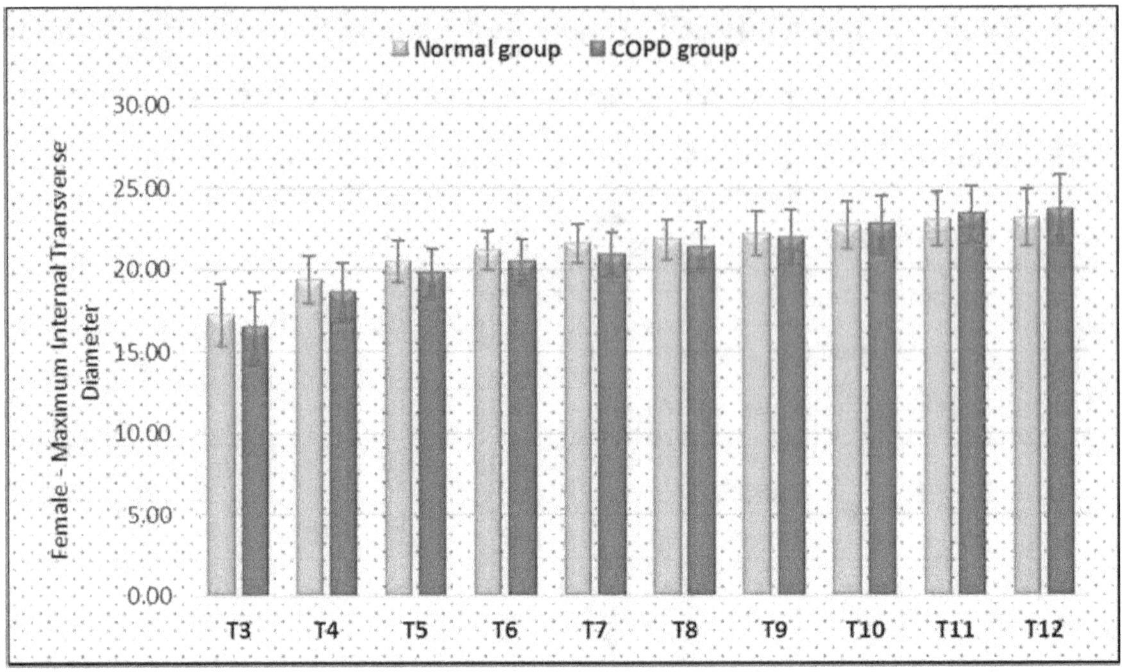

Graph No.- 1.06: Bar graph representing maximum transverse diameters per vertebral level in Normal and COPD groups among females.

- ❖ **Descriptive Summary and Comparison of Minimum Anteroposterior Diameter between Normal & COPD groups among Females (Table-8)**

The results-based descriptive summary and comparison of minimum anteroposterior diameter among females in the control and case groups using an unpaired T test illustrated that a significant association was found in both groups with T3 (t=-3.00; p<0.001),T4 (t=-10.54; p<0.001),T5 (t=-11.83; p<0.001), T6(t=-13.14; p<0.001), T7(t=-12.93; p<0.001), T8(t=-44.77; p<0.001), T9(t=-12.97; p<0.001), T10(t=-13.08; p<0.001), T11(t=-11.67; p<0.001) and T12(t=-10.92; p<0.001), (Table 8).

Table – 8: Descriptive Summary and Comparison of Minimum Anteroposterior Diameter between Normal & COPD groups among Females

Female - Minimum Anteroposterior Diameter	Normal group		COPD group		unpaired t test	
	Mean	SD	Mean	SD	t-value	p-value
T3	5.32	3.08	6.35	1.38	-3.00	**0.003**
T4	6.57	0.97	8.35	1.34	-10.54	**<0.001**
T5	7.80	1.02	9.83	1.34	-11.83	**<0.001**
T6	8.70	1.04	10.86	1.23	-13.14	**<0.001**
T7	9.22	1.14	11.43	1.23	-12.93	**<0.001**
T8	9.29	1.29	11.87	1.72	-44.77	**<0.001**
T9	9.49	1.35	12.17	1.50	-12.97	**<0.001**
T10	9.82	1.47	12.63	1.51	-13.08	**<0.001**
T11	10.08	1.59	12.85	1.70	-11.67	**<0.001**
T12	9.72	1.73	12.62	1.94	-10.92	**<0.001**

Table No. 8 demonstrated the shortest internal anteroposterior diameter at cranial level T3 and the longest at caudal level T11 in the normal and COPD groups among females. A regular increasing pattern was noted at each vertebral level from T3 to T11 and a decline at T12 as compared to T11. An interesting fact was also recorded: COPD females had higher values of anteroposterior diameters at each vertebral level from T3 to T12 as compared to normal females

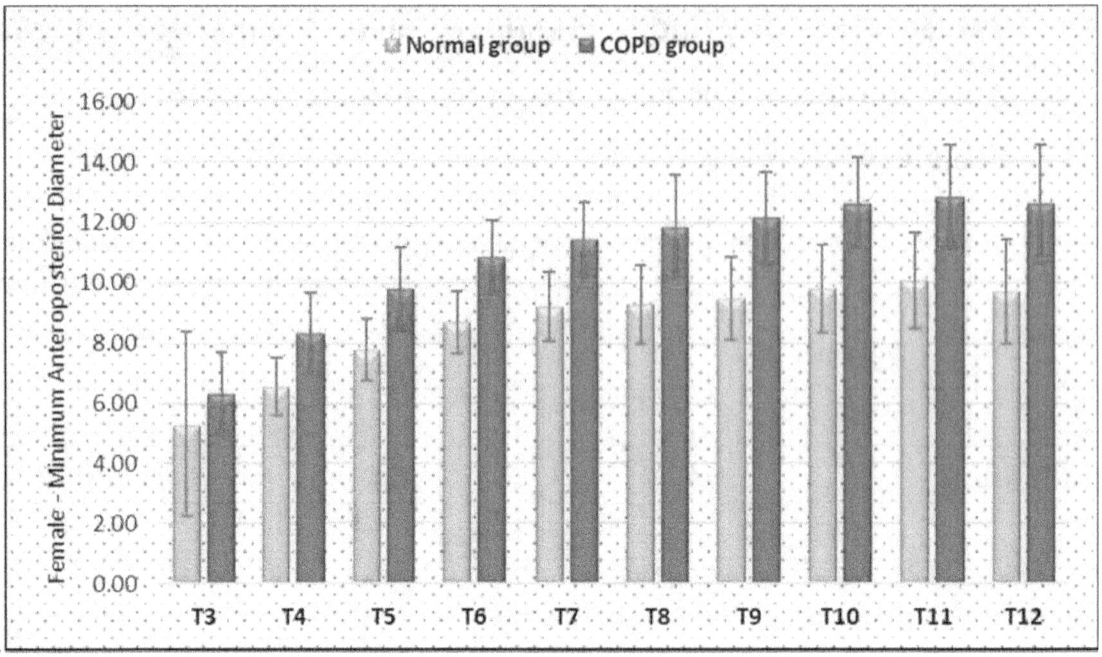

Graph No. 1.07: Bar graph showing minimum anteroposterior diameters and per vertebral levels in normal and COPD groups among females

❖ Descriptive Summary and Comparison of Haller Index between Normal & COPD groups among Females (Table-9)

The results-based descriptive summary and comparison of Haller index groups among female controls and cases using an unpaired T test revealed that there was a significant association in both the T3 and T4 groups ($t=9.67$; $p<0.001$), T4 ($t=13.48$; $p<0.001$), T5 ($t=12.87$; $p<0.001$), T6($t=12.88$; $p<0.001$), T7($t=11.72$; $p<0.001$), T8($t=8.53$; $p<0.001$), T9($t=7.62$; $p<0.001$), T10($t=9.12$; $p<0.001$), T11($t=8.62$; $p<0.001$) and T12($t=9.74$; $p<0.001$) (Table 9).

Table – 9: Descriptive Summary and Comparison of Haller Index between Normal & COPD groups among Females

Female - Haller Index	Normal group		COPD group		unpaired t test	
	Mean	SD	Mean	SD	t-value	p-value
T3	3.49	0.70	2.67	0.46	9.67	**<0.001**
T4	3.00	0.43	2.27	0.31	13.48	**<0.001**
T5	2.68	0.42	2.04	0.23	12.87	**<0.001**
T6	2.47	0.39	1.90	0.20	12.88	**<0.001**
T7	2.38	0.41	1.84	0.19	11.72	**<0.001**
T8	`2.39	0.62	1.82	0.21	8.53	**<0.001**
T9	2.42	0.73	1.83	0.21	7.62	**<0.001**
T10	2.38	0.57	1.82	0.19	9.12	**<0.001**
T11	2.35	0.55	1.84	0.20	8.62	**<0.001**
T12	2.45	0.50	1.90	0.25	9.74	**<0.001**

According to Table 9, females' mean Haller indices were highest at cranial level T3 in the control and COPD groups. In normal females, a decreasing pattern was observed from T3 to T11, except for T9, which was slightly higher than T8 and T10, and T12, which had a higher value as compared to T11. In COPD females, a decreasing pattern was observed from T3 to T8. Vertebral levels T9 had a slightly higher value as compared to T8 and T10, and the difference was insignificant. Vertebral level T11 had a higher value as compared to T10. Finally, an increment was noted at T12, which was a common pattern in both normal and COPD groups among females.

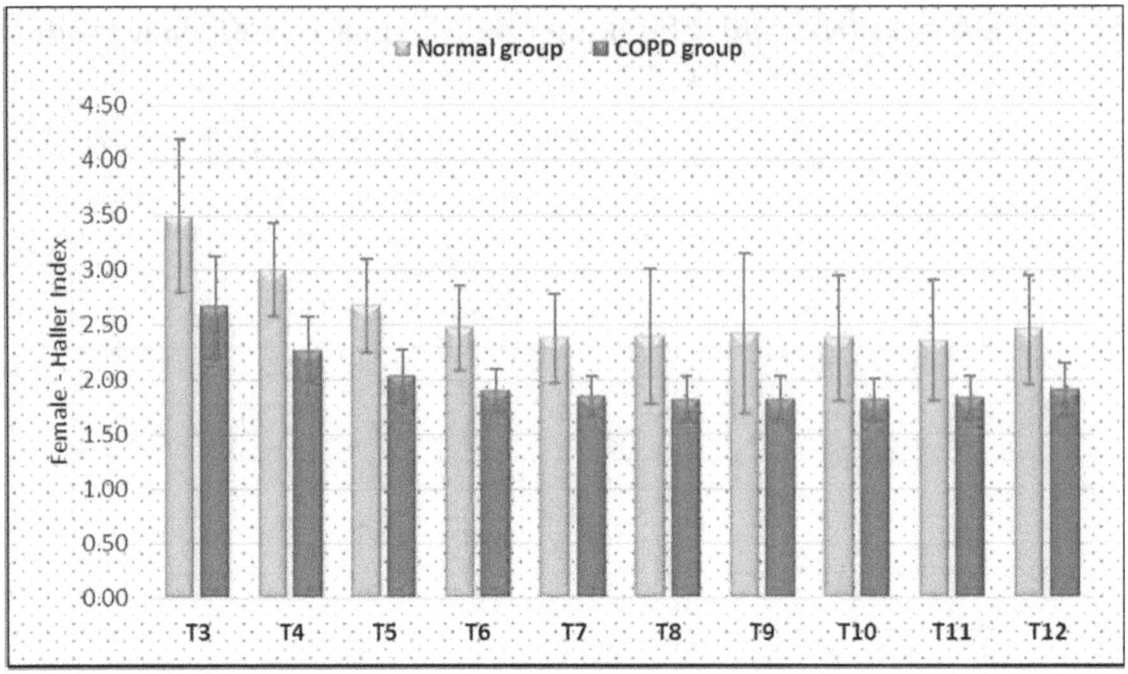

Graph No. - 1.08: Bar graph showing Haller indices per vertebral level in Normal and COPD group among females.

❖ Descriptive Summary and Comparison of Maximum Internal Transverse Diameter between Males & Females among Normal Group (Table-10)

The results-based descriptive summary and comparison of maximum internal transverse diameter among males and females in control and case groups using an unpaired ‘t’ test illustrated that a significant association was found in both groups with T3 (t=7.20; p<0.001),T4 (t=8.49; p<0.001),T5 (t=9.67; p<0.001), T6(t=10.38; p<0.001), T7(t=10.72; p<0.001), T8(t=10.65; p<0.001), T9(t=10.73; p<0.001), T10(t=10.60; p<0.001), T11(t=11.17; p<0.001) and T12(t=10.66; p<0.001) (Table 10).

Table – 10: Descriptive Summary and Comparison of Maximum Internal Transverse Diameter between Males & Females among Normal Group

Normal group - Maximum Internal Transverse Diameter	Male		Female		unpaired t test	
	Mean	SD	Mean	SD	t-value	p-value
T3	19.23	1.93	17.25	1.88	7.20	**<0.001**
T4	21.32	1.72	19.38	1.43	8.49	**<0.001**
T5	22.51	1.61	20.49	1.26	9.67	**<0.001**
T6	23.25	1.56	21.16	1.21	10.38	**<0.001**
T7	23.71	1.55	21.56	1.20	10.72	**<0.001**
T8	24.03	1.63	21.82	1.22	10.65	**<0.001**
T9	24.61	1.77	22.17	1.35	10.73	**<0.001**
T10	25.21	1.78	22.70	1.49	10.60	**<0.001**
T11	25.79	1.74	23.05	1.66	11.17	**<0.001**
T12	25.92	1.82	23.16	1.77	10.66	**<0.001**

Table No. 10 demonstrated that the maximum internal transverse diameter was smallest at cranial level T3 and highest at T12 in the normal group among males and females. A regular increment at each vertebral level, successively from T3 to T12, was noted in the normal group among males and females. It was also observed that females had smaller values of transverse diameters at each vertebral level as compared to males.

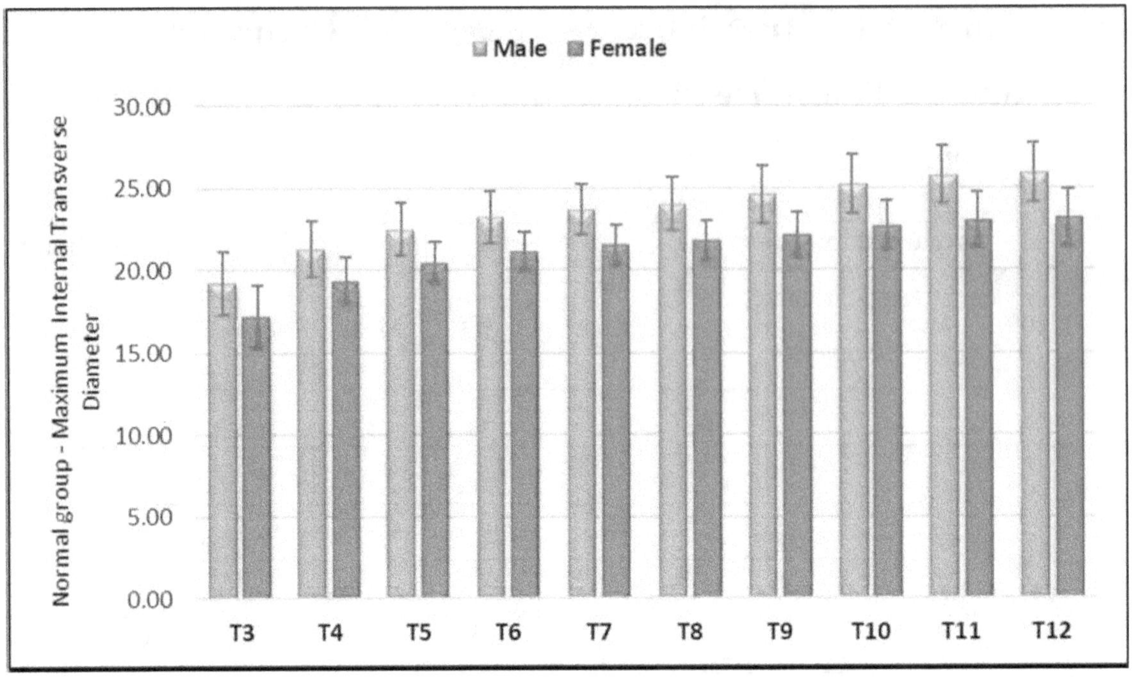

Graph No. - 1.09: Bar graph showing maximum internal diameters per vertebral level in Normal group among males and female.

- **Descriptive Summary and Comparison of Minimum Anteroposterior Diameter between Males & Females among Normal Group (Table-11)**

The result-based descriptive summary and comparison of minimum anteroposterior diameter among males and females in control and case groups using an unpaired T test illustrated that a significant association was found in both groups with T4 (t=4.77; p<0.001),T5 (t=4.73; p<0.001), T6(t=5.27; p<0.001), T7(t=5.90; p<0.001), T8(t=7.11; p<0.001), T9(t=6.33; p<0.001), T10(t=6.05; p<0.001), T11(t=6.32; p<0.001) andT12(t=6.06; p<0.001) (Table 11).

Table – 11: Descriptive Summary and Comparison of Minimum Anteroposterior Diameter between Males and Females among Normal Group

Normal group - Minimum Anteroposterior Diameter	Male		Female		unpaired t test	
	Mean	SD	Mean	SD	t-value	p-value
T3	5.57	0.89	5.32	3.08	0.78	0.437
T4	7.24	0.98	6.57	0.97	4.77	**<0.001**
T5	8.48	0.99	7.80	1.02	4.73	**<0.001**
T6	9.49	1.05	8.70	1.04	5.27	**<0.001**
T7	10.18	1.12	9.22	1.14	5.90	**<0.001**
T8	10.56	1.19	9.29	1.29	7.11	**<0.001**
T9	10.71	1.32	9.49	1.35	6.33	**<0.001**
T10	11.13	1.54	9.82	1.47	6.05	**<0.001**
T11	11.58	1.71	10.08	1.59	6.32	**<0.001**
T12	11.34	1.98	9.72	1.73	6.06	**<0.001**

According to Table No. 11, males and females in the Normal group had the smallest anteroposterior diameter at cranial level T3. A regular increasing pattern was observed from vertebral level T3 to T11 in both sexes in the normal group. Females had lower values of anteroposterior diameters at each vertebral level as compared to males. In both males and females, T12 had a lower value as compared to T11.

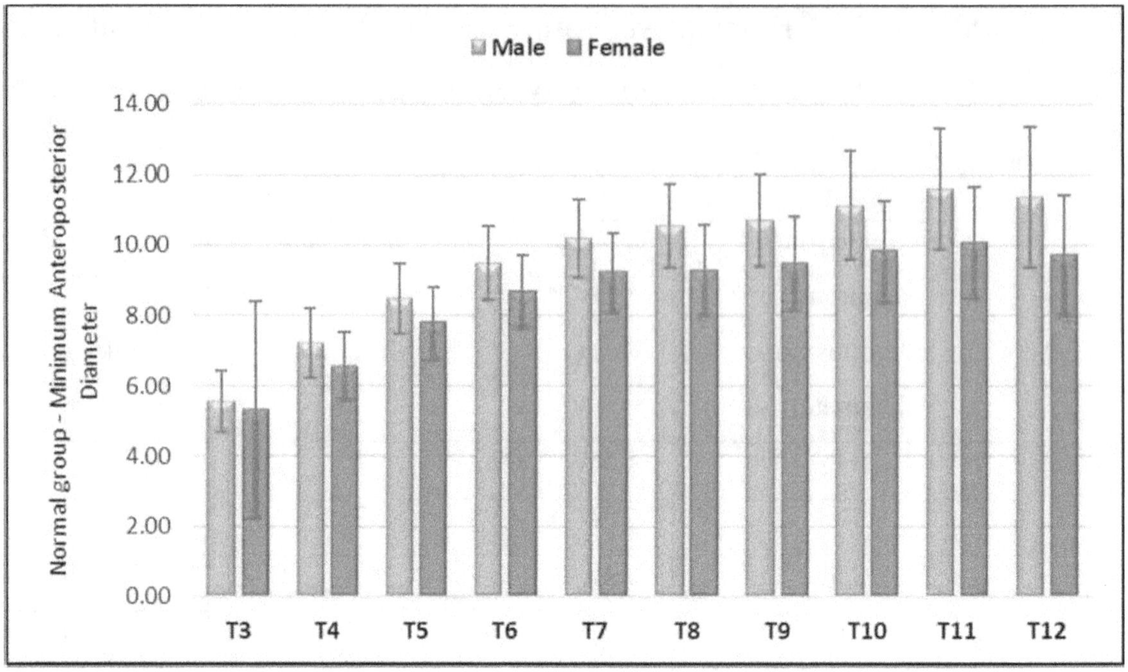

Graph No. - 1.10: Bar graph showing minimum anteroposterior per vertebral level in Normal group among males and females.

❖ Descriptive Summary and Comparison of Haller Index between Males & Females among Normal Group (Table-12)

The results-based descriptive summary and comparison of the Haller index among males and females in control and case groups using an unpaired T test illustrated that no significant association was found in any of the groups (Table 12).

Table – 12: Descriptive Summary and Comparison of Haller Indices between Males & Females among Normal Group

Normal group - Haller Index	Male		Female		unpaired t test	
	Mean	SD	Mean	SD	t-value	p-value
T3	3.52	0.53	3.49	0.70	0.28	0.777
T4	2.98	0.36	3.00	0.43	-0.40	0.690
T5	2.68	0.30	2.68	0.42	0.10	0.919
T6	2.47	0.26	2.47	0.39	0.02	0.983
T7	2.35	0.26	2.38	0.41	-0.61	0.543
T8	2.30	0.27	2.39	0.62	-1.35	0.180
T9	2.32	0.28	2.42	0.73	-1.15	0.253
T10	2.30	0.30	2.38	0.57	-1.21	0.229
T11	2.27	0.31	2.35	0.55	-1.38	0.169
T12	2.34	0.34	2.45	0.50	-1.91	0.058

Table 12 showed that Haller indices were highest in males and females in the normal group at cranial level T3. In males, Haller indices were observed in a decreasing pattern from T3 to T11, while T9 and T10 had equal values and a slightly raised value at T9. Vertebral level T12 had a higher magnitude as compared to T11. In females, Haller indices were observed in decreasing order from T3 to T7. Further, with slight increments at T8 and T9, then reduced values at T10 and T11. Finally, T12 has a higher value than T11. Females had higher values of the Haller indices at the per vertebral level as compared to males.

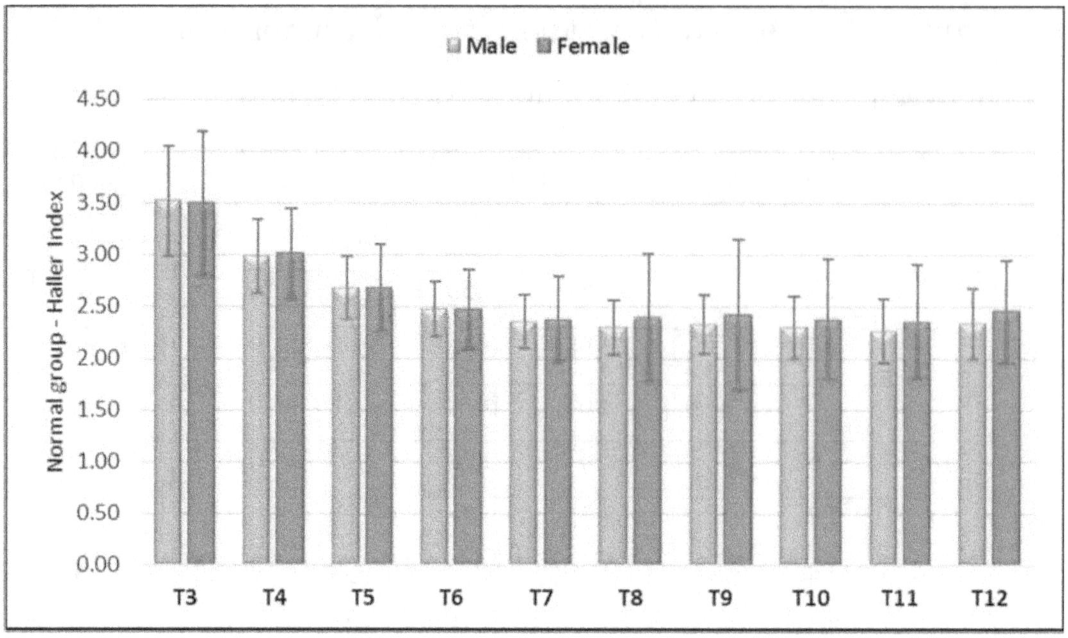

Graph No. - 1.11: Bar graph showing Haller indices per vertebral level in males and females among Normal group (Higher in females except T3).

- **Descriptive Summary and Comparison of Maximum Internal Transverse Diameter between Males & Females among COPD Group (Table-13)**

The descriptive summary of the results and comparison of maximum internal transverse diameter among COPD group among males and females in control and cases using an unpaired T test revealed that a significant association with T3 was found in both groups ($t=7.57$; $p<0.001$),T4 ($t=9.39$; $p<0.001$),T5 ($t=11.43$; $p<0.001$), T6($t=12.37$; $p<0.001$), T7($t=12.25$; $p<0.001$), T8($t=9.52$; $p<0.001$), T9($t=11.25$; $p<0.001$), T10($t=10.65$; $p<0.001$), T11($t=10.81$; $p<0.001$) and T12($t=8.76$; $p<0.001$) (Table 13).

Table – 13: Descriptive Summary and Comparison of Maximum Internal Transverse Diameter between Males & Females among COPD Group

COPD group - Maximum Internal Transverse Diameter	Male		Female		unpaired t test	
	Mean	SD	Mean	SD	t-value	p-value
T3	18.69	1.89	16.46	2.17	7.57	**<0.001**
T4	20.85	1.47	18.64	1.77	9.39	**<0.001**
T5	22.05	1.23	19.79	1.50	11.43	**<0.001**
T6	22.82	1.25	20.47	1.39	12.37	**<0.001**
T7	23.31	1.31	20.91	1.40	12.25	**<0.001**
T8	23.70	1.89	21.36	1.50	9.52	**<0.001**
T9	24.64	1.66	21.97	1.63	11.25	**<0.001**
T10	25.50	1.82	22.75	1.76	10.65	**<0.001**
T11	26.16	1.85	23.35	1.75	10.81	**<0.001**
T12	26.39	2.20	23.64	2.16	8.76	**<0.001**

Table No. 13 demonstrated that the maximum internal transverse diameter was smallest at cranial level T3 and longest at T12 in males and females among the COPD group. An increasing pattern was observed from proximal vertebral level T3 to distal vertebral level T12. It was also observed that transverse diameters per vertebral level were higher in males as compared to females in the COPD group.

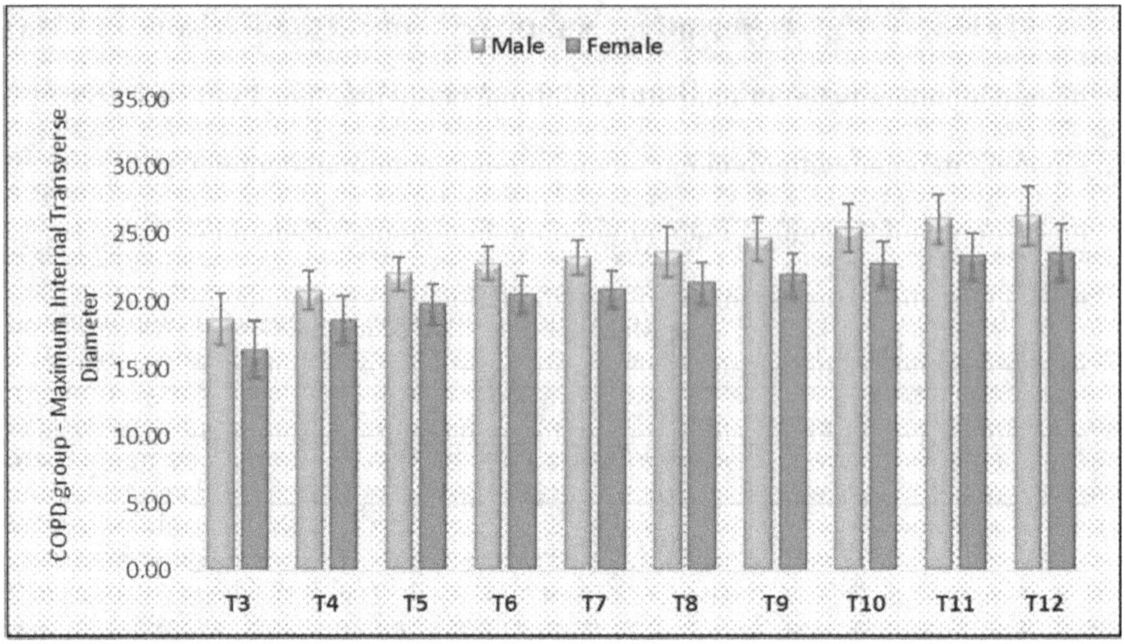

Graph No. - 1.12: Bar graph showing transverse diameter per vertebral level in males and females among COPD group.

- **Descriptive Summary and Comparison of Minimum Anteroposterior Diameter between Males & Females among COPD Group (Table-14)**

The results-based descriptive summary and comparison of minimum anteroposterior diameter among males and females in control and case groups using an unpaired T test illustrated that a significant association was found in both groups with T3 (t=3.50; p<0.001),T4 (t=3.64; p<0.001),T5 (t=4.61; p<0.001), T6(t=5.88; p<0.001), T7(t=7.51; p<0.001), T8(t=6.24; p<0.001), T9(t=6.69; p<0.001), T10(t=6.84; p<0.001), T11(t=6.65; p<0.001) and T12(t=5.01; p<0.001) (Table 14).

Table – 14: Descriptive Summary and Comparison of Minimum Anteroposterior Diameter between Males & Females among COPD Group

COPD group - Minimum Anteroposterior Diameter	Male		Female		unpaired t test	
	Mean	SD	Mean	SD	t-value	p-value
T3	7.02	1.30	6.35	1.38	3.50	**0.001**
T4	9.07	1.43	8.35	1.34	3.64	**<0.001**
T5	10.71	1.31	9.83	1.34	4.61	**<0.001**
T6	11.92	1.26	10.86	1.23	5.88	**<0.001**
T7	12.77	1.23	11.43	1.23	7.51	**<0.001**
T8	13.28	1.37	11.87	1.72	6.24	**<0.001**
T9	13.74	1.74	12.17	1.50	6.69	**<0.001**
T10	14.36	1.97	12.63	1.51	6.84	**<0.001**
T11	14.67	2.08	12.85	1.70	6.65	**<0.001**
T12	14.22	2.45	12.62	1.94	5.01	**<0.001**

Table No. 14 demonstrated that in COPD males and females, the anteroposterior diameter was found to be smallest at cranial vertebral level T3 and highest at T11, with a slightly increased value at T12. A consistently increasing pattern was observed at each successive vertebral level from T3 to T11, with a decreased value at T12 compared to T11. COPD females had a smaller anteroposterior diameter as compared to COPD males.

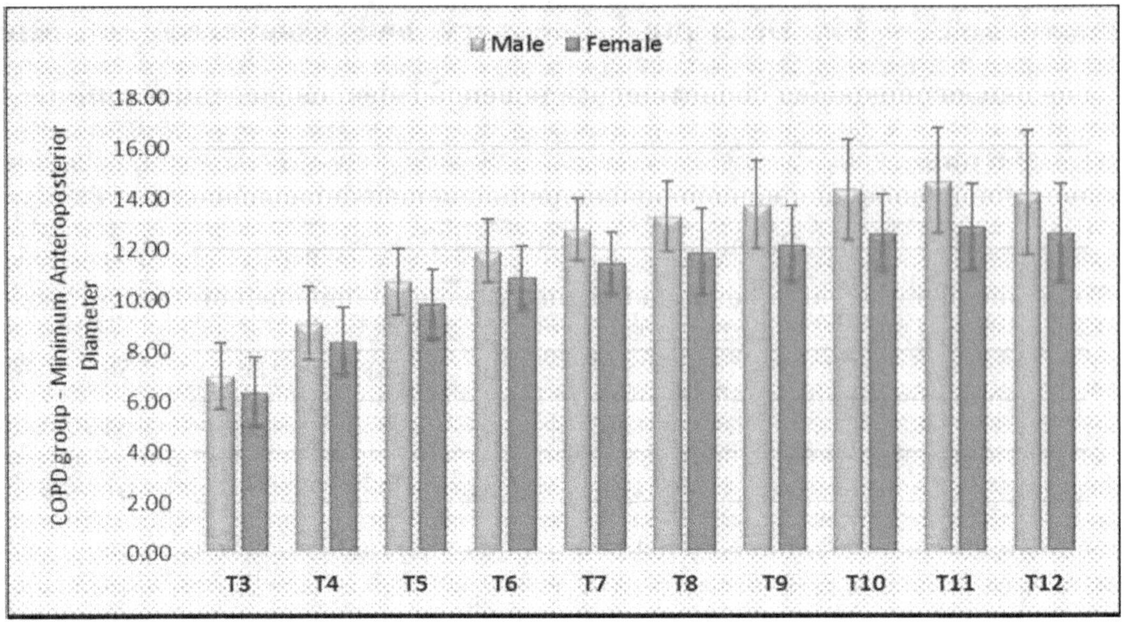

Graph No. - 1.13: Bar graph showing anteroposterior diameter per vertebral level in males and females among COPD groups.

❖ Descriptive Summary and Comparison of Haller Index between Males & Females among COPD Group (Table-15)

The result-based descriptive summary and comparison of Haller Index male and female control and case samples using an unpaired T test illustrated that no significant association was found in any of the groups (Table 15).

Table – 15: Descriptive Summary and Comparison of Haller Indices between Males & Females among COPD Group

COPD group - Haller Index	Male		Female		unpaired t test	
	Mean	SD	Mean	SD	t-value	p-value
T3	2.72	0.39	2.67	0.46	0.82	0.412
T4	2.34	0.30	2.27	0.31	1.52	0.131
T5	2.08	0.23	2.04	0.23	1.31	0.192
T6	1.93	0.18	1.90	0.20	1.06	0.290
T7	1.84	0.16	1.84	0.19	-0.26	0.796
T8	1.80	0.19	1.82	0.21	-0.84	0.402
T9	1.81	0.19	1.83	0.21	-0.42	0.677
T10	1.80	0.19	1.82	0.19	-0.75	0.452
T11	1.81	0.19	1.84	0.20	-1.13	0.261
T12	1.90	0.28	1.90	0.25	-0.11	0.912

Table 15 showed that the Haller index was highest in males and females with COPD at cranial vertebral level T3. In males, a decreasing pattern was observed from vertebral level T3 to T8, with a slight increase in value at T9 and T11. Finally, an increased value was observed at T12 as compared to T11. In COPD females, a decreasing pattern was observed from T3 to T8, then a slight increment at T9 and T11. Ultimately, vertebral level T12 had a higher value than T11. In females, Haller indices per vertebral level were smaller as compared to males.

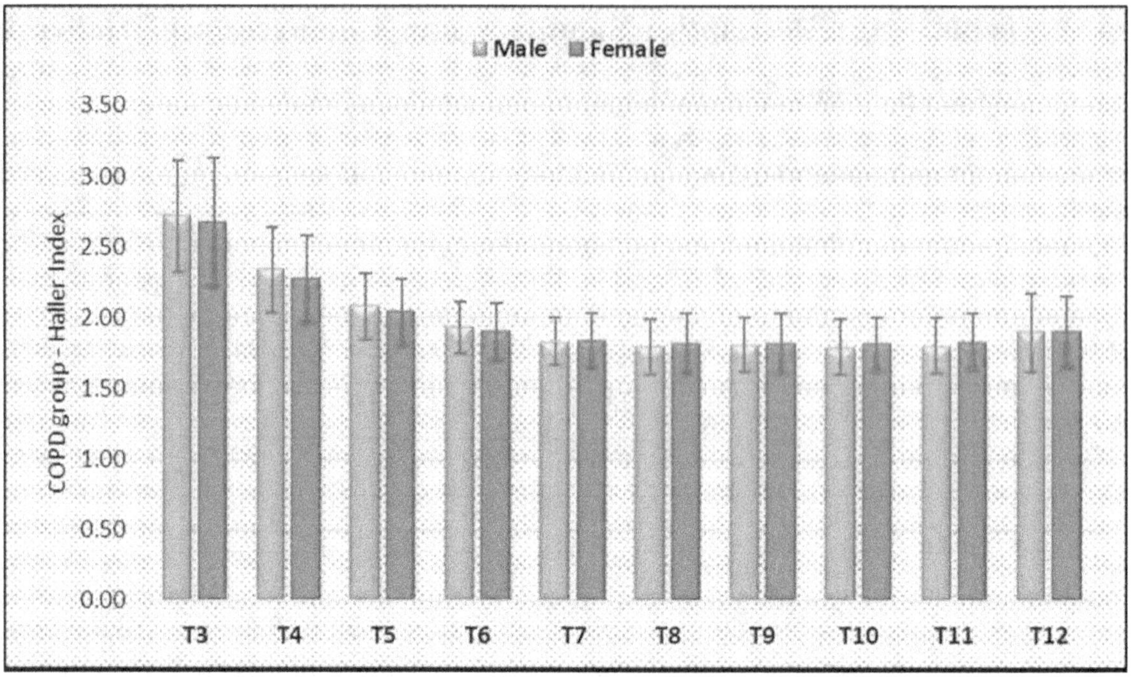

Graph No. - 1.14: Bar graph showing Haller Indices per vertebral in males and females among COPD groups.

- **Descriptive Summary and Comparison of Maximum Internal Transverse Diameter among Various Age Groups among Normal Group (Table-16)**

The results-based descriptive summary and comparison of maximum internal transverse diameter among various age groups in the normal group using ANOVA illustrated that a significant association was found in all the groups 18-29 yr (21.41±1.85);30 - 39 yr (21. 78±1.86);40 - 49 yr(21.54±1.43);50 - 59 yr (20.77±1.66);60-70 yr (22.11±1.77) with T5 (p=0.019) and 18 - 29 yr (22.11±1.90);30-39yr (22.60±1.65);40-49 yr (22. 26 ± 1 . 58) ;50-59 yr (21.49±1.68); 60 - 70 yr(22.65±1.67) with T6(p=0.026) (Table 16).

Table – 16: Descriptive Summary and Comparison of Maximum Internal Transverse Diameter among Various Age Groups among Normal Group

Normal group - Maximum Internal Transverse Diameter	18 - 29 yr		30 - 39 yr		40 - 49 yr		50 - 59 yr		60 - 70 yr		ANOVA	
	Mean	SD	Mean	SD	Mean	SD	Mean	SD	Mean	SD	F-value	p-value
T3	18.23	2.28	18.69	2.19	18.15	2.04	17.57	2.16	18.61	1.89	1.66	0.161
T4	20.27	2.10	20.72	1.78	20.39	1.54	19.69	1.89	20.75	1.68	2.07	0.086
T5	21.41	1.85	21.78	1.86	21.54	1.43	20.77	1.66	22.11	1.77	3.03	**0.019**
T6	22.11	1.90	22.60	1.65	22.26	1.58	21.49	1.68	22.65	1.67	2.84	**0.026**
T7	22.49	1.86	23.01	1.78	22.71	1.58	22.00	1.73	23.04	1.64	2.32	0.058
T8	22.77	1.79	23.27	1.88	22.99	1.77	22.37	1.83	23.31	1.70	1.77	0.136
T9	23.14	2.00	23.70	2.05	23.72	2.03	22.83	1.94	23.69	1.81	1.65	0.163
T10	23.59	2.05	24.37	2.10	24.17	1.98	23.45	2.08	24.35	2.04	1.76	0.140
T11	23.91	2.12	24.89	2.26	24.59	2.01	23.99	2.21	24.91	2.19	1.98	0.099
T12	23.91	2.02	24.86	2.44	24.67	2.08	24.27	2.32	25.20	2.36	2.01	0.095

Table No. 16 demonstrated that the mean transverse diameter at each vertebral level showed a slight increment in the age group of 30-39 years as compared to 18-29 years. Age groups 40-49 and 50-59 years had slightly lower values of transverse diameters at each vertebral level than age groups 30-39 years. Except for the 30-39-year group, the age group 60-70 years had a higher value of transverse diameter per vertebral level.

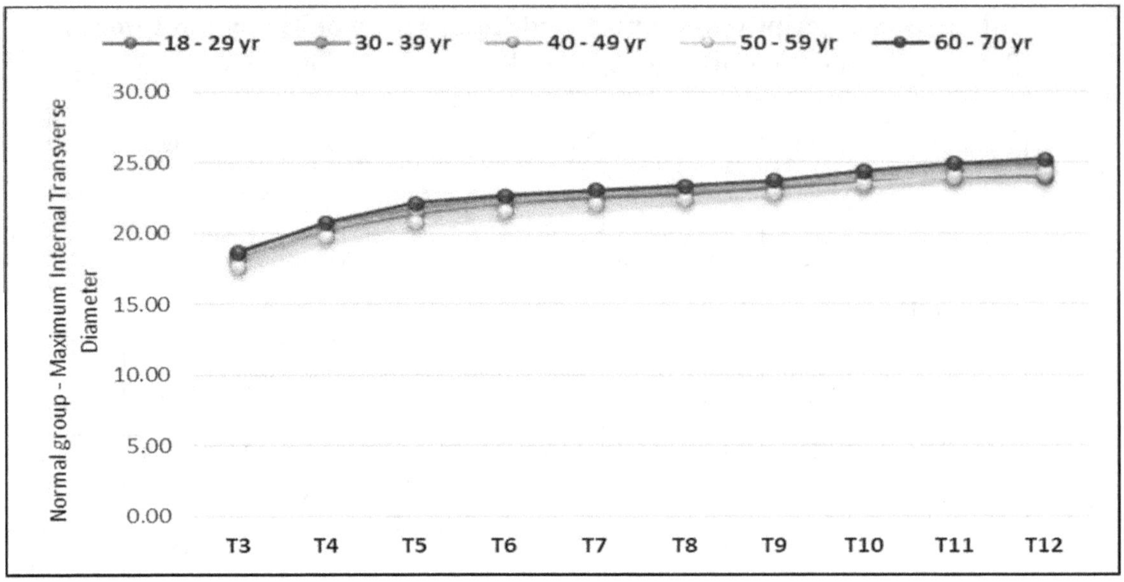

Graph No. - 1.15: Linear graph showing transverse diameter per vertebral level in various age groups among control group.

- ### ❖ Descriptive Summary and Comparison of Minimum Antero-posterior Diameter among Various Age Groups among Normal Group (Table-17)

The results-based descriptive summary and comparison of minimum anteroposterior diameter among various age groups in the normal group using ANOVA illustrated that no significant association was found in any of the groups (Table 17).

Table – 17: Descriptive Summary and Comparison of Minimum Antero-posterior Diameter among Various Age Groups among Normal Group

Normal group - Minimum Anteroposterior Diameter	18 - 29 yr		30 - 39 yr		40 - 49 yr		50 - 59 yr		60 - 70 yr		ANOVA	
	Mean	SD	Mean	SD	Mean	SD	Mean	SD	Mean	SD	F-value	p-value
T3	5.65	4.34	5.41	0.98	5.37	0.91	5.09	0.98	5.69	0.86	0.44	0.778
T4	6.67	1.06	6.97	0.95	6.97	1.13	6.75	0.98	7.27	0.96	1.98	0.100
T5	7.97	1.06	8.21	1.03	8.17	1.22	8.00	1.03	8.43	0.96	1.12	0.348
T6	8.95	1.05	9.16	1.08	9.06	1.40	8.97	1.08	9.40	0.93	0.95	0.434
T7	9.57	1.15	9.73	1.14	9.68	1.56	9.58	1.20	10.00	1.06	0.72	0.580
T8	9.87	1.32	10.02	1.32	9.92	1.76	9.80	1.28	10.04	1.33	0.19	0.942
T9	10.03	1.29	10.32	1.43	9.99	1.93	9.96	1.46	10.23	1.20	0.46	0.764
T10	10.27	1.44	10.70	1.77	10.47	1.98	10.45	1.57	10.52	1.47	0.35	0.841
T11	10.48	1.54	11.07	1.90	10.98	2.11	10.84	1.79	10.87	1.79	0.65	0.628
T12	10.12	1.64	10.63	2.11	10.85	2.39	10.60	1.90	10.57	2.16	0.74	0.568

Table No. 17 demonstrated that anteroposterior diameters were increased at the per vertebral level in the age group 30-39 years as compared to 18-29 years, except for T3, which had a lower value as compared to the age group 18-29. Then, decreased values of anteroposterior diameter were observed at the vertebral level in the age groups 40–49 and 50–59 years as compared to 30-39 years. The 60-70-year age group had a larger anteroposterior diameter and per vertebral level than the other age groups. T11 and T12 vertebral levels were highest in the age group 40–49 years, compared to other age groups in normal subjects. T9 and T10 vertebral levels were highest in the 30-39-year age group when compared to other age groups in the normal group. There was no significant difference or association in either group

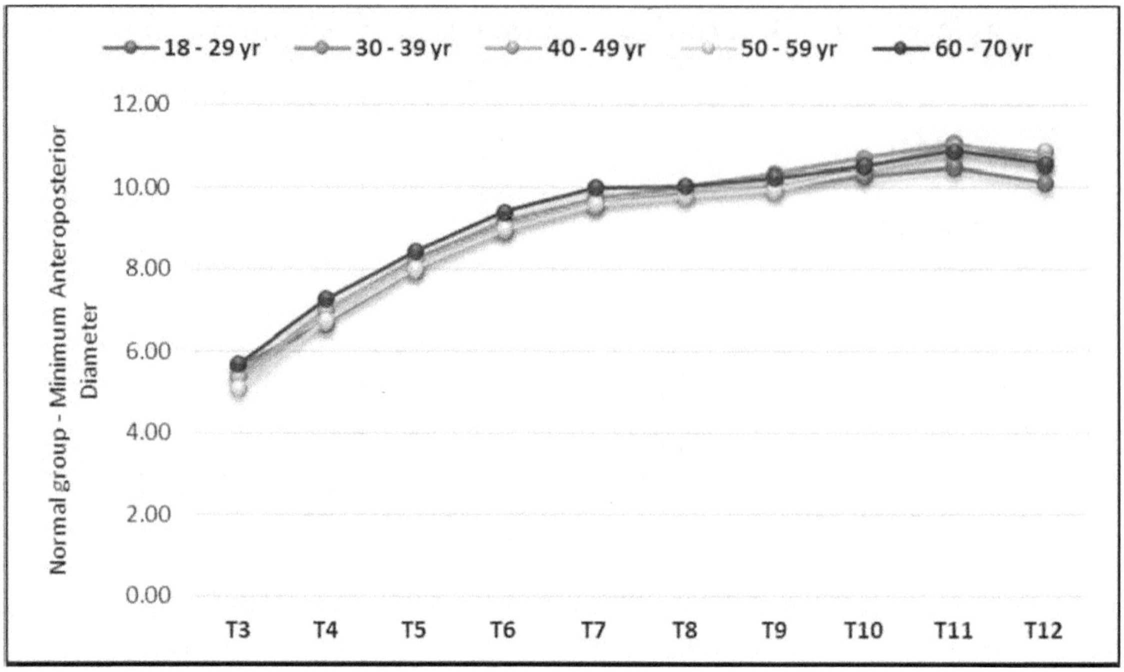

Graph No. - 1.16: Linear graph showing anteroposterior diameter per vertebral level in various age groups among Normal group.

- **Descriptive Summary and Comparison of Haller Index among Various Age Groups among Normal Group (Table-18)**

The results-based summary and comparison of the Haller index in various age groups among the normal group using ANOVA illustrated that no significant association was found in any of the groups (Table 18).

Table – 18: Descriptive Summary and Comparison of Haller Index among Various Age Groups among Normal Group

Normal group - Haller Index	18 - 29 yr		30 - 39 yr		40 - 49 yr		50 - 59 yr		60 - 70 yr		ANOVA	
	Mean	SD	Mean	SD	Mean	SD	Mean	SD	Mean	SD	F-value	p-value
T3	3.62	0.72	3.53	0.59	3.47	0.72	3.53	0.52	3.31	0.44	1.23	0.298
T4	3.08	0.38	3.01	0.34	3.00	0.55	2.96	0.35	2.89	0.32	1.30	0.271
T5	2.72	0.31	2.68	0.33	2.70	0.56	2.63	0.30	2.65	0.31	0.44	0.783
T6	2.49	0.25	2.49	0.26	2.53	0.56	2.42	0.27	2.43	0.23	0.73	0.573
T7	2.37	0.25	2.39	0.25	2.43	0.58	2.32	0.28	2.32	0.26	0.61	0.659
T8	2.34	0.34	2.35	0.25	2.37	0.92	2.31	0.26	2.36	0.36	0.08	0.987
T9	2.33	0.29	2.32	0.23	2.55	1.15	2.33	0.32	2.34	0.30	1.12	0.348
T10	2.33	0.28	2.32	0.27	2.44	0.87	2.29	0.34	2.35	0.31	0.58	0.675
T11	2.31	0.28	2.29	0.28	2.37	0.85	2.25	0.31	2.34	0.35	0.36	0.838
T12	2.40	0.30	2.40	0.35	2.40	0.71	2.34	0.32	2.45	0.40	0.34	0.852

Table No. 18 The Haller indices per vertebral level decreased from T3 to T5 across various age groups. At T5 vertebral level, alternate increasing and decreasing patterns were observed across the age groups. At vertebral level, T6 Heller indices had equal values in age groups 18–29 and 30-39, then increased in 40–49 as compared to the previous two groups, and again had less value in age groups 50–59 and 60–70. T7 Haller indices showed an increasing pattern from age groups 18-29 to 40-49, with age groups 50-59 and 60-70 having the same value. At vertebral level, T8 Haller indices were showing an increasing pattern from age groups 18–29 to 40–49. The value was lower in the age group 50-59 years than in the age group 40-49 years, but slightly higher in the age group 60-70 years. The T9 Haller Index was lower in the 30-39 age group compared to the 18-29 age group at the vertebral level. While values in the age groups 50-59 and 60-70 were lower than in the 40-49 age group, at vertebral level T10, alternate increases and decreases in the value of Haller indices were noted across the age groups. At vertebral level T11, the same pattern was observed as at vertebral level T10. At vertebral level T12, equal values of Haller indices were observed in the age groups 18–29, 30-39, and 40–49 years. While the age group 50-59 had lower values for per vertebral level than the age group 40-49, the age group 60-70 had the highest value compared to the other age groups.

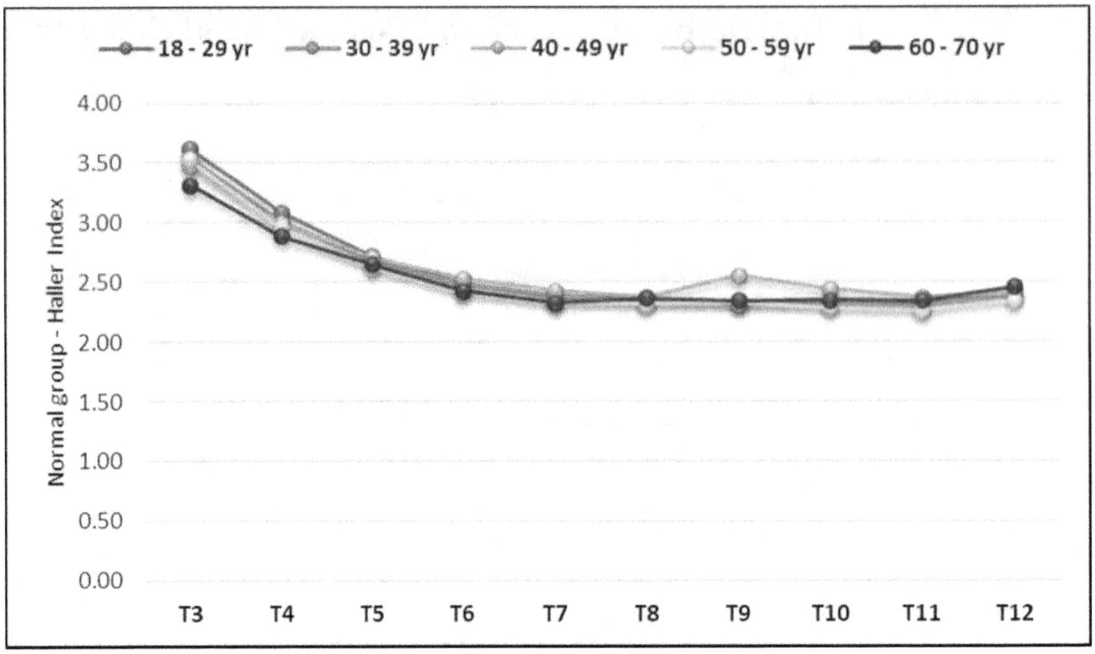

Graph No. - 1.17: Linear graph showing Haller Indices per-vertebral level in various age groups among Normal group.

- ### Descriptive Summary and Comparison of Maximum Internal Transverse Diameter among Various Age Groups among COPD Group (Table-19)

The results-based descriptive summary and comparison of maximum internal transverse diameter among various age groups in the COPD group using ANOVA illustrated that a significant association was found in all the groups 18 - 29 yr (21.16±2.99); 30 - 39 yr (23.28±2.34); 40 - 49 yr (21.54±1.79); 50 - 59 yr (22.80±1.74); 60 - 70 yr (22.45±1.89) with T8 (p=0.014) and 18 - 29 yr (22.75±2.45); 30 - 39 yr (24.76±2.66); 40-49 yr (24.08±1.97);50-59 yr (24.42±2.07), 60 - 70 yr (24.12±2.22) with T10 (p=0.044), 18 - 29 yr (23.14±2.62); 30 - 39 yr (25.37±2.66); 40 - 49 yr (24.62±2.02); 50 - 59 yr(25.09±2.00);60 - 70 yr (24.83±2.23) with T11 (p=0.013) 18 - 29 yr (23.18±2.073); 30-39 yr (25.66±2.68); 40 - 49 yr (24.79±2.00); 50 - 59 yr (25.48±2.00); 60 - 70 yr (25.08±2.87) with T12 (p=0.010) (Table 19).

Table – 19: Descriptive Summary and Comparison of Maximum Internal Transverse Diameter among Various Age Groups among COPD Group

COPD group - Maximum Internal Transverse Diameter	18 - 29 yr		30 - 39 yr		40 - 49 yr		50 - 59 yr		60 - 70 yr		ANOVA	
	Mean	SD	Mean	SD	Mean	SD	Mean	SD	Mean	SD	F-value	p-value
T3	17.14	2.30	17.59	2.27	17.42	2.97	17.67	2.02	17.69	2.24	0.26	0.901
T4	19.32	2.25	19.98	1.70	19.56	2.45	19.87	1.69	19.78	1.91	0.43	0.786
T5	20.40	2.26	21.26	1.71	20.86	1.81	21.00	1.61	20.93	1.76	0.65	0.631
T6	21.04	2.21	22.08	1.77	21.59	1.73	21.78	1.62	21.60	1.76	0.98	0.419
T7	21.37	2.14	22.69	1.96	22.06	1.67	22.32	1.63	22.00	1.83	1.61	0.173
T8	21.16	2.99	23.28	2.34	22.56	1.79	22.80	1.74	22.45	1.89	3.20	**0.014**
T9	22.32	2.42	24.02	2.60	23.33	1.96	23.51	1.82	23.19	2.08	1.84	0.122
T10	22.75	2.45	24.76	2.66	24.08	1.97	24.42	2.07	24.12	2.22	2.50	**0.044**
T11	23.14	2.62	25.37	2.66	24.62	2.02	25.09	2.00	24.83	2.23	3.26	**0.013**
T12	23.18	2.73	25.66	2.68	24.79	2.14	25.48	2.00	25.08	2.87	3.40	**0.010**

Table No. 19 showed an increasing pattern across the age group from the cranial to the distal level. When horizontal tracing was observed at each vertebral level, usually increased values were observed as we moved across the higher age groups.

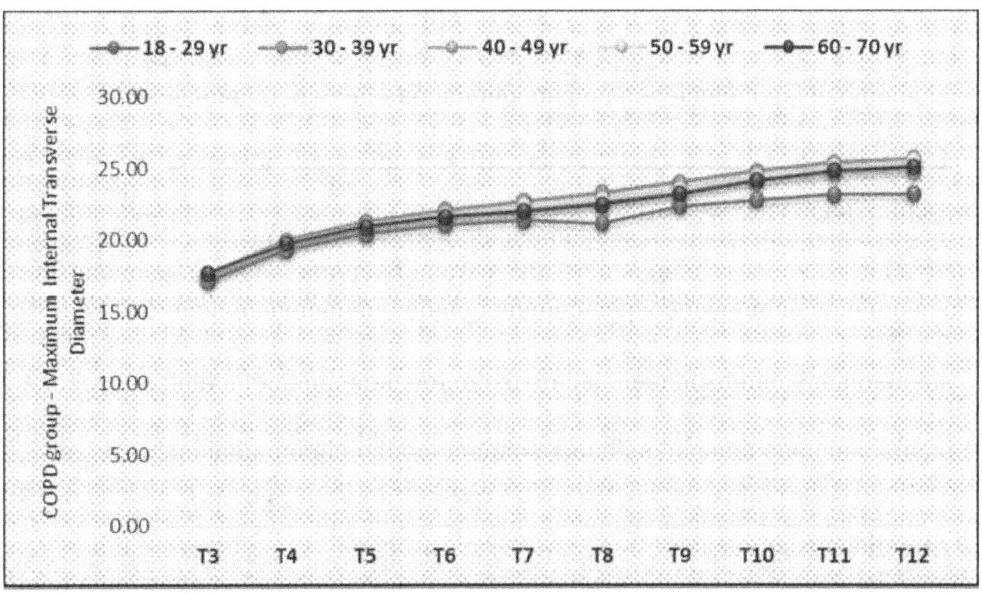

Graph No. - 1.18: Linear graph showing maximum internal transverse diameter per-vertebral level across the various age groups among COPD group.

❖ Descriptive Summary and Comparison of Minimum Antero-posterior Diameter among Various Age Groups among COPD Group (Table-20)

The result-based summary and comparison of minimum anteroposterior diameter among various age groups in the COPD group using ANOVA illustrated that no significant association was found in any of the groups (Table 20).

Table – 20: Descriptive Summary and Comparison of Minimum Antero-posterior Diameter among Various Age Groups among COPD Group

COPD group - Minimum Antero-posterior Diameter	18 - 29 yr		30 - 39 yr		40 - 49 yr		50 - 59 yr		60 - 70 yr		ANOVA	
	Mean	SD	Mean	SD	Mean	SD	Mean	SD	Mean	SD	F-value	p-value
T3	6.32	0.96	6.58	1.13	6.64	1.70	6.66	1.48	6.87	1.31	0.68	0.609
T4	8.56	1.16	8.32	1.22	8.78	1.54	8.60	1.47	8.92	1.47	0.95	0.439
T5	9.95	1.22	10.12	1.22	10.33	1.43	10.10	1.44	10.50	1.44	0.96	0.432
T6	11.00	1.34	11.11	1.23	11.51	1.46	11.27	1.36	11.62	1.32	1.27	0.284
T7	11.56	1.37	11.82	1.39	12.21	1.39	12.01	1.39	12.35	1.40	1.59	0.178
T8	11.84	1.43	12.49	1.80	12.56	1.53	12.77	2.09	12.67	1.48	1.12	0.350
T9	12.14	1.38	12.97	1.98	13.03	2.01	12.99	1.72	13.12	1.79	1.14	0.340
T10	12.46	1.46	13.52	2.27	13.63	2.23	13.70	1.80	13.55	1.91	1.53	0.194
T11	12.41	1.61	14.01	2.28	13.85	2.43	13.97	1.90	13.87	2.05	2.27	0.064
T12	11.96	1.69	13.89	2.41	13.48	2.57	13.75	1.94	13.39	2.54	2.39	0.052

Table No. 20 showed that COPD patients had the smallest antero-posterior diameter at cranial level T3 and the largest at T11, with a slightly lower value at the T12 per-vertebral level. An increasing pattern was observed at vertebral levels T3–T11 in all age groups. When horizontal tracing was made at each vertebral level across the age groups, an increasing pattern was seen across the age groups. Antero-posterior diameter was found to be greater in the 60-70 age group compared to the 18-29 age group. At vertebral level T10, an increasing pattern was seen up to the age group 50–59 and a slightly lower value in the age group 60–70. At vertebral level T11, a horizontally alternate increasing and decreasing pattern was observed. At vertebral level T12, alternately increased and decreased values were observed. On average, increased values of anteroposterior diameter were observed across the age groups.

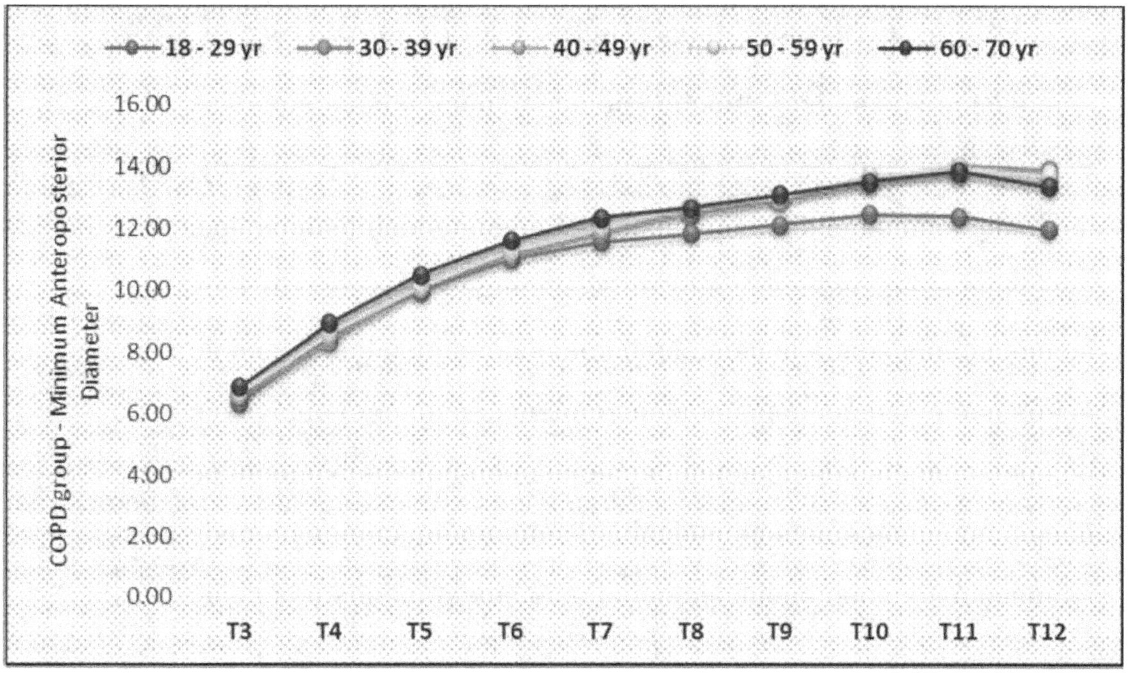

Graph No. - 1.19: Linear graph showing anteroposterior diameter per-vertebral level in various age groups among COPD group.

❖ Descriptive Summary and Comparison of Haller Index among Various Age Groups among COPD Group (Table-21)

The results-based descriptive summary and comparison of the Haller index among various age groups in the COPD group using ANOVA illustrated that a significant association was found in all the groups 18 - 29 yr (1.86±0.14);30 - 39 yr (1.93±0.13); 40 - 49 yr(1.82±0.16); 50 - 59 yr(1.87±0.18); 60 - 70 yr (1.80±0.19) with T7 (p=0.014) (Table 21).

Table – 21: Descriptive Summary and Comparison of Haller Index in Various Age Groups among COPD Group

COPD group - Haller Index	18 - 29 yr		30 - 39 yr		40 - 49 yr		50 - 59 yr		60 - 70 yr		ANOVA	
	Mean	SD	Mean	SD	Mean	SD	Mean	SD	Mean	SD	F-value	p-value
T3	2.74	0.29	2.72	0.42	2.69	0.42	2.74	0.48	2.64	0.42	0.55	0.699
T4	2.27	0.21	2.44	0.34	2.26	0.27	2.36	0.32	2.26	0.32	2.13	0.079
T5	2.06	0.17	2.12	0.21	2.04	0.20	2.11	0.25	2.02	0.25	1.43	0.225
T6	1.92	0.16	2.00	0.14	1.89	0.17	1.95	0.19	1.88	0.20	2.39	0.052
T7	1.86	0.14	1.93	0.13	1.82	0.16	1.87	0.18	1.80	0.19	3.11	**0.016**
T8	1.80	0.25	1.88	0.15	1.81	0.16	1.82	0.24	1.79	0.20	0.86	0.486
T9	1.85	0.18	1.87	0.15	1.82	0.20	1.83	0.21	1.79	0.21	0.83	0.506
T10	1.83	0.16	1.85	0.16	1.79	0.20	1.80	0.20	1.80	0.20	0.44	0.777
T11	1.87	0.18	1.83	0.16	1.81	0.23	1.82	0.20	1.81	0.19	0.43	0.790
T12	1.95	0.20	1.87	0.19	1.88	0.25	1.88	0.22	1.92	0.33	0.50	0.739

Table no. 21 showed the highest Haller Index values were observed at cranial level T3, with subsequent decreasing patterns observed up to vertebral level T11 on average, with minor variations, and finally higher values at vertebral level T12 as compared to T11. The per vertebral level was higher in the 30-39 age group compared to 18-29 years.

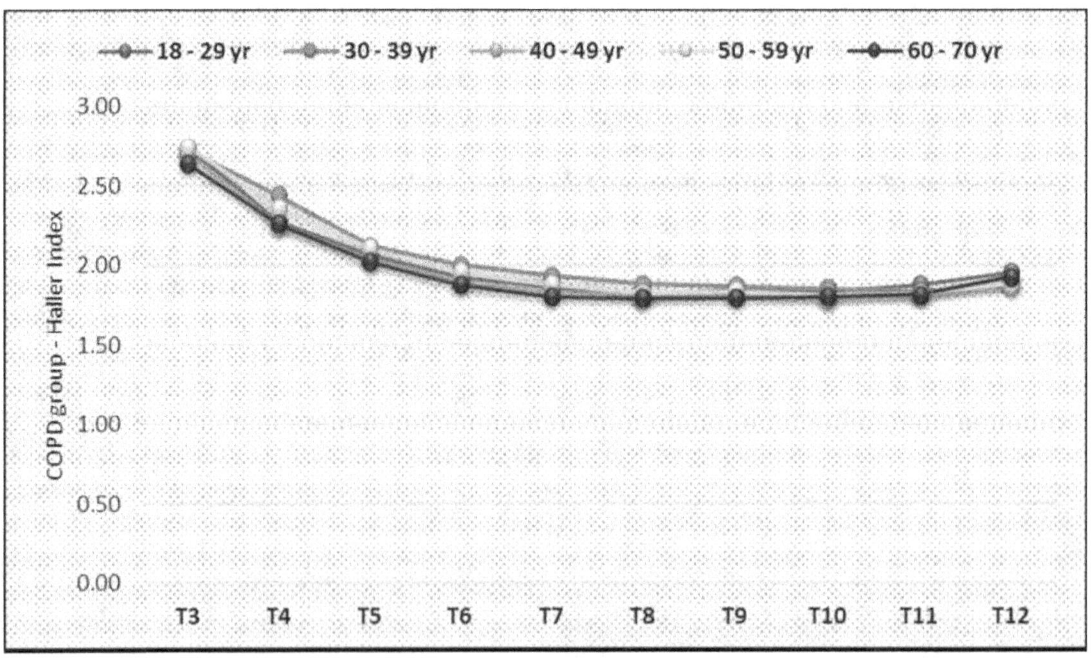

Graph No. - 1.20: Linear graph showing Haller index per-vertebral level Various Age Groups among COPD Group

❖ Descriptive Summary and Comparison of Maximum Internal Transverse Diameter between Normal and COPD Groups among age group 18 – 29 years (Table-22)

The results-based summary and comparison of maximum internal transverse diameter between normal and COPD groups among age groups 18-29 yr for control and cases using an unpaired T test revealed that a significant association was found in both the group with T7 and the group without T7 (t=2.12; p=0.038) and T9(t=2.69; p=0.009) (Table 22).

Table – 22: Descriptive Summary and Comparison of Maximum Internal Transverse Diameter between Normal and COPD Groups among age group 18 – 29 years.

18 - 29 yr - Maximum Internal Transverse Diameter	Normal group		COPD group		unpaired t test	
	Mean	SD	Mean	SD	t-value	p-value
T3	18.23	2.28	17.14	2.30	1.74	0.086
T4	20.27	2.10	19.32	2.25	1.63	0.109
T5	21.41	1.85	20.40	2.26	1.89	0.064
T6	22.11	1.90	21.04	2.21	1.98	0.052
T7	22.49	1.86	21.37	2.14	2.12	**0.038**
T8	22.77	1.79	21.16	2.99	2.69	**0.009**
T9	23.14	2.00	22.32	2.42	1.41	0.165
T10	23.59	2.05	22.75	2.45	1.43	0.159
T11	23.91	2.12	23.14	2.62	1.24	0.218
T12	23.91	2.02	23.18	2.73	1.19	0.239

Table No. 22 demonstrated that the maximum internal transverse diameter was smallest at cranial vertebral level T3 in the normal and COPD groups. An increasing pattern was observed from vertebral level T3 to T12 in both groups. When compared to the Control group, the COPD group had lower values at the per-vertebral level.

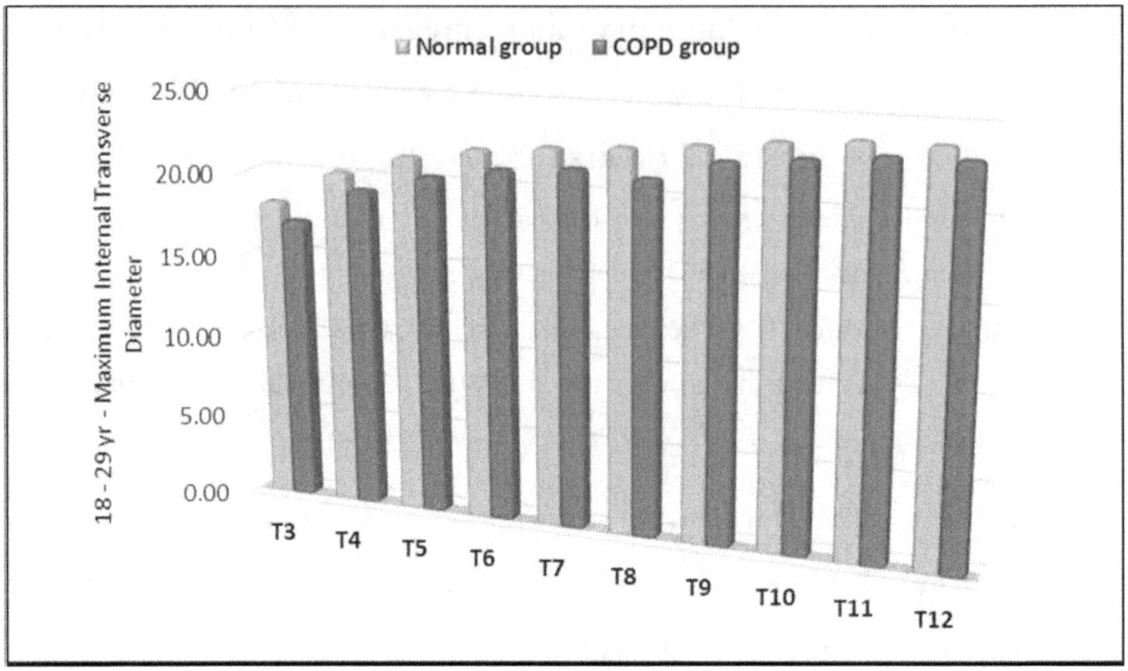

Graph No. - 1.21: Bar graph showing transverse diameter per-vertebral level among Normal and COPD groups in age group 18-29 years.

- ### Descriptive Summary and Comparison of Minimum Anteroposterior Diameter between Normal and COPD Groups among age group 18 – 29 years (Table-23)

The findings, based on a descriptive summary and comparison of minimum anteroposterior diameter between normal and COPD groups in the age group 18-29 years, revealed a significant association in all groups with T3 ($t=-0.66$; $p<0.001$),T4 ($t=-6.36$; $p<0.001$),T5 ($t=-6.60$; $p<0.001$), T6($t=-6.00$; $p<0.001$), T7($t=-5.35$; $p<0.001$), T8($t=-5.84$; $p<0.001$), T9($t=-5.54$; $p<0.001$), T10($t=-5.54$; $p<0.001$), T11($t=-4.54$; $p<0.001$) and T12($t=-4.10$; $p<0.001$) (Table 23).

Table – 23: Descriptive Summary and Comparison of Minimum Anteroposterior Diameter between Normal and COPD Groups among age group 18 – 29 years.

18 - 29 yr - Minimum Anteroposterior Diameter	Normal group		COPD group		unpaired t test	
	Mean	SD	Mean	SD	t-value	p-value
T3	5.65	4.34	6.32	0.96	-0.66	0.513
T4	6.67	1.06	8.56	1.16	-6.36	**<0.001**
T5	7.97	1.06	9.95	1.22	-6.55	**<0.001**
T6	8.95	1.05	11.00	1.34	-6.60	**<0.001**
T7	9.57	1.15	11.56	1.37	-6.00	**<0.001**
T8	9.87	1.32	11.84	1.43	-5.35	**<0.001**
T9	10.03	1.29	12.14	1.38	-5.84	**<0.001**
T10	10.27	1.44	12.46	1.46	-5.54	**<0.001**
T11	10.48	1.54	12.41	1.61	-4.54	**<0.001**
T12	10.12	1.64	11.96	1.69	-4.10	**<0.001**

Table No. 23 demonstrated that the minimum anteroposterior diameter was smallest at cranial vertebral level T3 in the normal and COPD groups. An increasing pattern was observed from vertebral level T3 to T11, with a reduced value at vertebral level T12 as compared to vertebral level T11 in both groups. In the age group of 18-29 years, the COPD group had higher values per vertebral level than the control group.

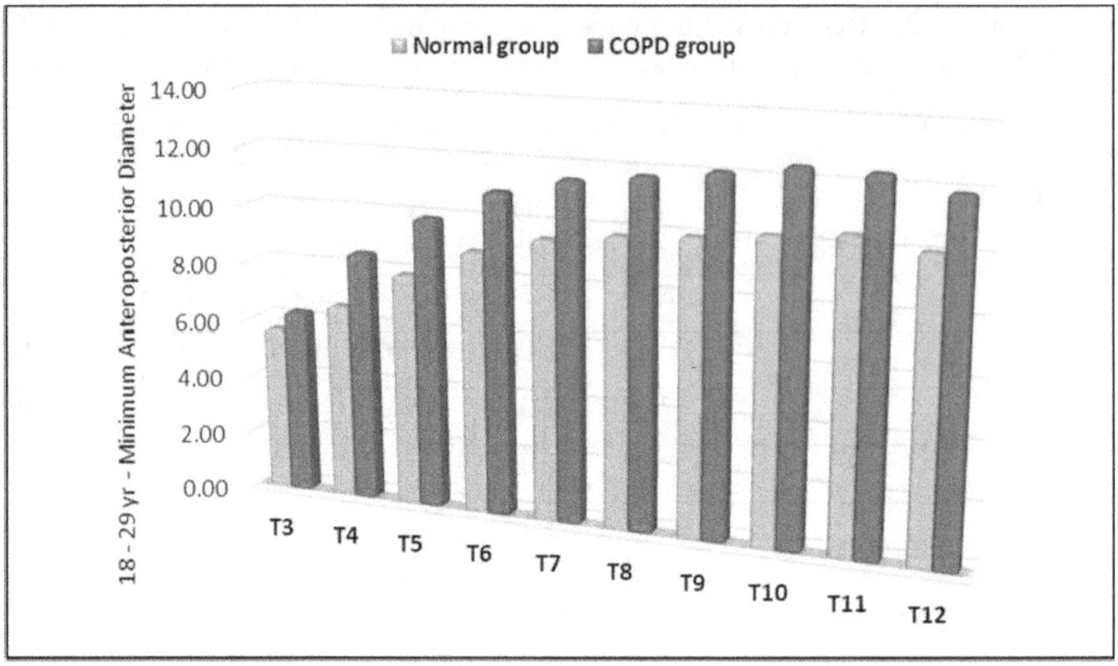

Graph No. - 1.22: Bar graph showing minimum anteroposterior diameter per-vertebral level in age group 18-29 years among Normal and COPD groups.

- **Descriptive Summary and Comparison of Haller Index between Normal and COPD groups among age group 18–29 years (Table-24)**

The descriptive summary and comparison of the Haller index between normal and COPD groups among the age group of 18-29 years using an unpaired t test revealed a significant association in all groups with T3 ($t=5.14$; $p<0.001$),T4 ($t=8.74$; $p<0.001$),T5 ($t=8.66$; $p<0.001$), T6($t=9.19$; $p<0.001$), T7($t=8.51$; $p<0.001$), T8($t=6.33$; $p<0.001$), T9($t=6.79$; $p<0.001$), T10($t=7.21$; $p<0.001$), T11($t=6.30$; $p<0.001$) and T12($t=5.99$; $p<0.001$) (Table 24).

Table – 24: Descriptive Summary and Comparison of Haller Index between Normal and COPD Groups among age group 18 – 29 years.

18 - 29 yr - Haller Index	Normal group		COPD group		unpaired t test	
	Mean	SD	Mean	SD	t-value	p-value
T3	3.62	0.72	2.74	0.29	5.14	**<0.001**
T4	3.08	0.38	2.27	0.21	8.74	**<0.001**
T5	2.72	0.31	2.06	0.17	8.66	**<0.001**
T6	2.49	0.25	1.92	0.16	9.19	**<0.001**
T7	2.37	0.25	1.86	0.14	8.51	**<0.001**
T8	2.34	0.34	1.80	0.25	6.33	**<0.001**
T9	2.33	0.29	1.85	0.18	6.79	**<0.001**
T10	2.33	0.28	1.83	0.16	7.21	**<0.001**
T11	2.31	0.28	1.87	0.18	6.30	**<0.001**
T12	2.40	0.30	1.95	0.20	5.99	**<0.001**

Table No. 24 demonstrated that the Haller index was highest at cranial vertebral T3 in the age group of 18–29 years among the normal and COPD groups. Haller indices per vertebral level were seen in a decreasing pattern from cranial T3 to T11 and an increment at T12 as compared to T11 among the normal and COPD groups in the 18–29 age group. The COPD group's Haller index per vertebral was found to be lower than the normal group.

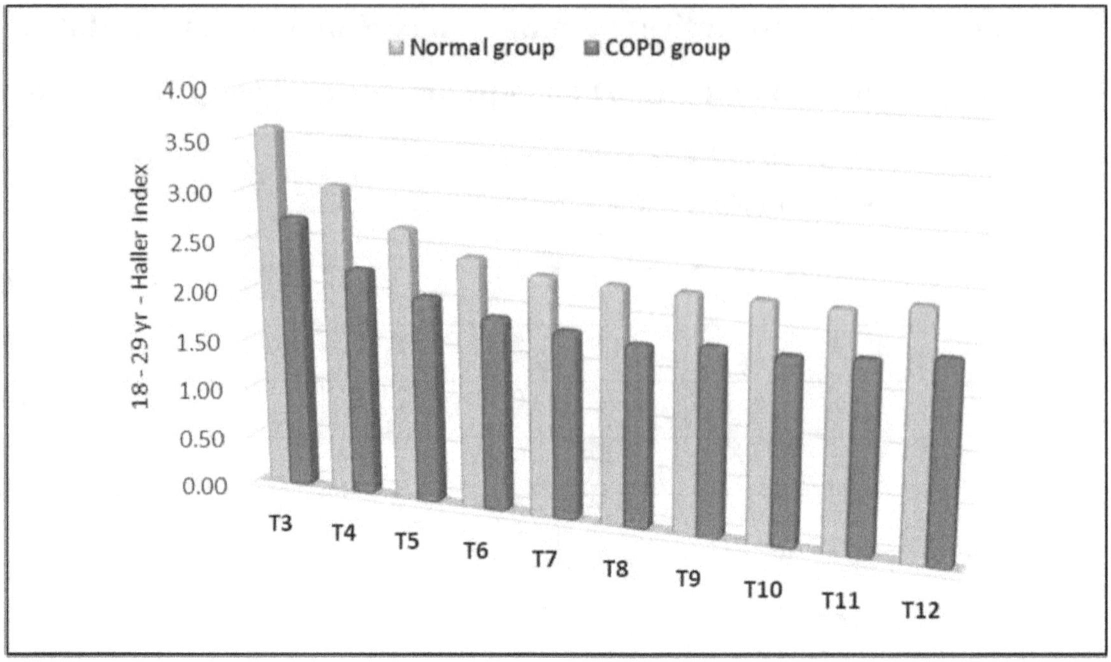

Graph No. - 1.23: Bar graph showing Haller index per-vertebral level in age group 18-29 years anong Normal and COPD groups.

❖ **Descriptive Summary and Comparison of Maximum Internal Transverse Diameter between Normal and COPD Groups among age group 30 – 39 years (Table-25)**

The descriptive summary and comparison of maximum internal transverse diameter between the normal and COPD groups among the 30-39-year age group using an unpaired t test revealed that no significant association was found in any of the groups (Table 25).

Table – 25: Descriptive Summary and Comparison of Maximum Internal Transverse Diameter between Normal and COPD Groups among age group 30 – 39 years.

<table>
<tr><th rowspan="2">30 - 39 yr - Maximum Internal Transverse Diameter</th><th colspan="2">Normal group</th><th colspan="2">COPD group</th><th colspan="2">unpaired t test</th></tr>
<tr><th>Mean</th><th>SD</th><th>Mean</th><th>SD</th><th>t-value</th><th>p-value</th></tr>
<tr><td>T3</td><td>18.69</td><td>2.19</td><td>17.59</td><td>2.27</td><td>1.85</td><td>0.069</td></tr>
<tr><td>T4</td><td>20.72</td><td>1.78</td><td>19.98</td><td>1.70</td><td>1.58</td><td>0.119</td></tr>
<tr><td>T5</td><td>21.78</td><td>1.86</td><td>21.26</td><td>1.71</td><td>1.08</td><td>0.284</td></tr>
<tr><td>T6</td><td>22.60</td><td>1.65</td><td>22.08</td><td>1.77</td><td>1.16</td><td>0.251</td></tr>
<tr><td>T7</td><td>23.01</td><td>1.78</td><td>22.69</td><td>1.96</td><td>0.66</td><td>0.512</td></tr>
<tr><td>T8</td><td>23.27</td><td>1.88</td><td>23.28</td><td>2.34</td><td>-0.01</td><td>0.989</td></tr>
<tr><td>T9</td><td>23.70</td><td>2.05</td><td>24.02</td><td>2.60</td><td>-0.53</td><td>0.598</td></tr>
<tr><td>T10</td><td>24.37</td><td>2.10</td><td>24.76</td><td>2.66</td><td>-0.63</td><td>0.530</td></tr>
<tr><td>T11</td><td>24.89</td><td>2.26</td><td>25.37</td><td>2.66</td><td>-0.76</td><td>0.453</td></tr>
<tr><td>T12</td><td>24.86</td><td>2.44</td><td>25.66</td><td>2.68</td><td>-1.19</td><td>0.238</td></tr>
</table>

Table No. 25 demonstrated the smallest transverse diameter at proximal vertebral level T3 and the highest value at T12. Transverse diameters at each vertebra were observed in an increasing trend from T3 to T12. Transverse vertebral diameters from T3 to T7 were lower in the COPD group among 30-39-year olds. While a reverse pattern was observed from vertebral T8 to T12, the COPD group had higher values of transverse diameter per-vertebral level as compared to the normal group in the 30-39 age group. The difference between the upper and lower thoracic segments.

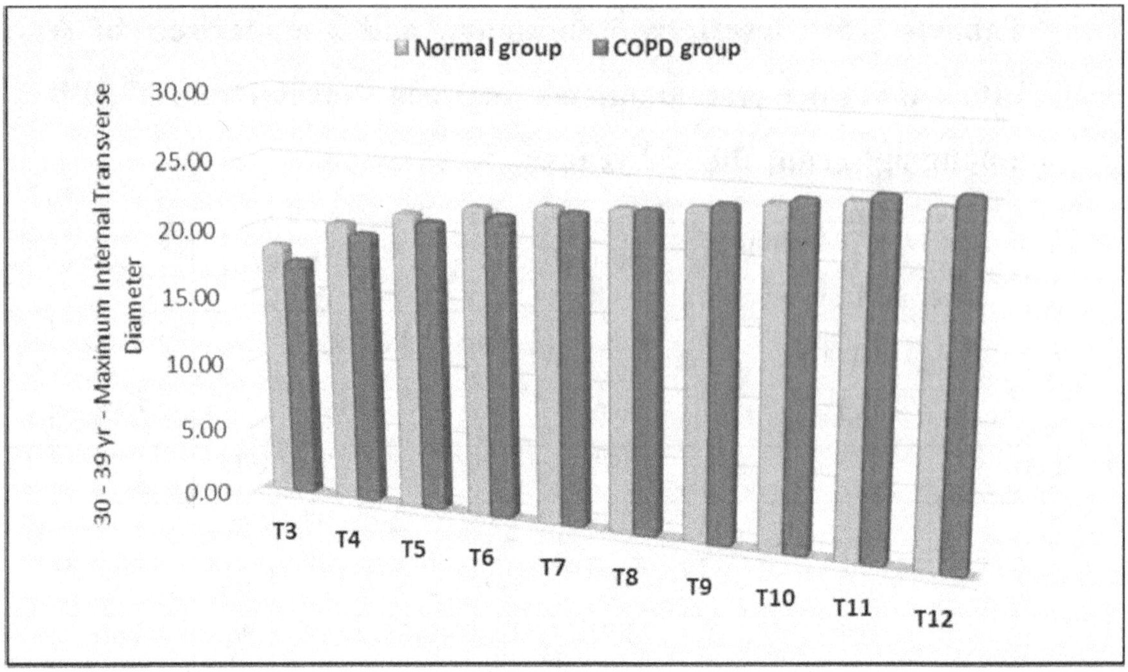

Graph No. - 1.24: Bar graph showing maximum internal transverse diameter per-vertebral level in age group 30-39 among Normal and COPD groups.

- **Descriptive Summary and Comparison of Minimum Anteroposterior Diameter between Normal and COPD Groups among age group 30 – 39 years (Table-26)**

The results of a descriptive summary and comparison of minimum anteroposterior diameter between normal and COPD groups in the age group 30-39 yr using an unpaired t test revealed that a significant association with T3 was found in all groups (t=-4.24; p<0.001),T4 (t=-4.78; p<0.001),T5 (t=-6.51; p<0.001), T6(t=-6.44; p<0.001), T7(t=-6.37; p<0.001), T8(t=-6.13; p<0.001), T9(t=-6.02; p<0.001), T10(t=-5.41; p<0.001), T11(t=-5.39; p<0.001) and T12(t=-5.50; p<0.001) (Table 26).

Table – 26: Descriptive Summary and Comparison of Minimum Anteroposterior Diameter between Normal and COPD Groups among age group 30 – 39 yr

30 - 39 yr - Minimum Anteroposterior Diameter	Normal group		COPD group		unpaired t test	
	Mean	SD	Mean	SD	t-value	p-value
T3	5.41	0.98	6.58	1.13	-4.24	**<0.001**
T4	6.97	0.95	8.32	1.22	-4.78	**<0.001**
T5	8.21	1.03	10.12	1.22	-6.51	**<0.001**
T6	9.16	1.08	11.11	1.23	-6.44	**<0.001**
T7	9.73	1.14	11.82	1.39	-6.37	**<0.001**
T8	10.02	1.32	12.49	1.80	-6.13	**<0.001**
T9	10.32	1.43	12.97	1.98	-6.02	**<0.001**
T10	10.70	1.77	13.52	2.27	-5.41	**<0.001**
T11	11.07	1.90	14.01	2.28	-5.39	**<0.001**
T12	10.63	2.11	13.89	2.41	-5.50	**<0.001**

Table No. 26 demonstrated that the minimum anteroposterior diameter was smallest at proximal vertebral T3 and largest at T11. Anteroposterior diameters were observed in an increasing pattern from vertebral level T3 to T11 and a slightly decreased value at T12 as compared to T11 in the age group 30-39 years among the normal and COPD groups. Anteroposterior diameter at the per-vertebral level was observed higher in the COPD group as compared to the normal group.

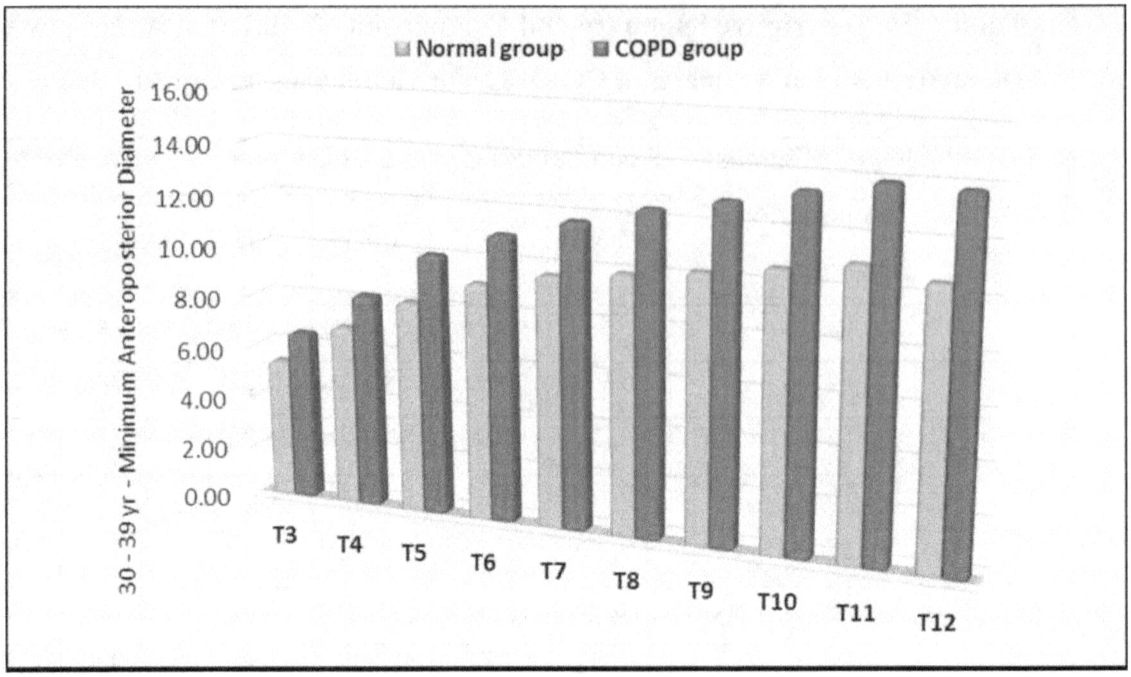

Graph No. -1.25: Bar graph showing minimum anteroposterior diameter per-vertebral level in age group 30-39 years among Normal and COPD group.

- **Descriptive Summary and Comparison of Haller Index between Normal and COPD Groups among age group 30 – 39 years (Table 27)**

The descriptive summary and comparison of the Haller index between normal and COPD groups in the age group 30-39 years using an unpaired t test revealed a significant association in all groups with T3 (t=5.74; p<0.001),T4 (t=6.27; p<0.001),T5 (t=7.22; p<0.001), T6(t=8.15; p<0.001), T7(t=8.02; p<0.001), T8(t=8.02; p<0.001), T9(t=8.19; p<0.001), T10(t=7.27; p<0.001), T11(t=7.01; p<0.001) and T12(t=6.48; p<0.001) (Table 27).

Table – 27: Descriptive Summary and Comparison of Haller Index between Normal and COPD Groups among age group 30 – 39 yr

30 - 39 yr - Haller Index	Normal group		COPD group		unpaired t test	
	Mean	SD	Mean	SD	t-value	p-value
T3	3.53	0.59	2.72	0.42	5.74	**<0.001**
T4	3.01	0.34	2.44	0.34	6.27	**<0.001**
T5	2.68	0.33	2.12	0.21	7.22	**<0.001**
T6	2.49	0.26	2.00	0.14	8.15	**<0.001**
T7	2.39	0.25	1.93	0.13	8.02	**<0.001**
T8	2.35	0.25	1.88	0.15	8.02	**<0.001**
T9	2.32	0.23	1.87	0.15	8.19	**<0.001**
T10	2.32	0.27	1.85	0.16	7.27	**<0.001**
T11	2.29	0.28	1.83	0.16	7.01	**<0.001**
T12	2.40	0.35	1.87	0.19	6.48	**<0.001**

Table No. 27 demonstrated that the Haller index was highest at cranial vertebral T3 in the age group 30-39 years among the normal and COPD groups. Haller indices per vertebral level were seen in a decreasing pattern from cranial T3 to T11 and an increment at T12 as compared to T11 among the normal and COPD groups in the 30-39 age group. The COPD group's Haller indices per vertebral were found to be lower than the normal group.

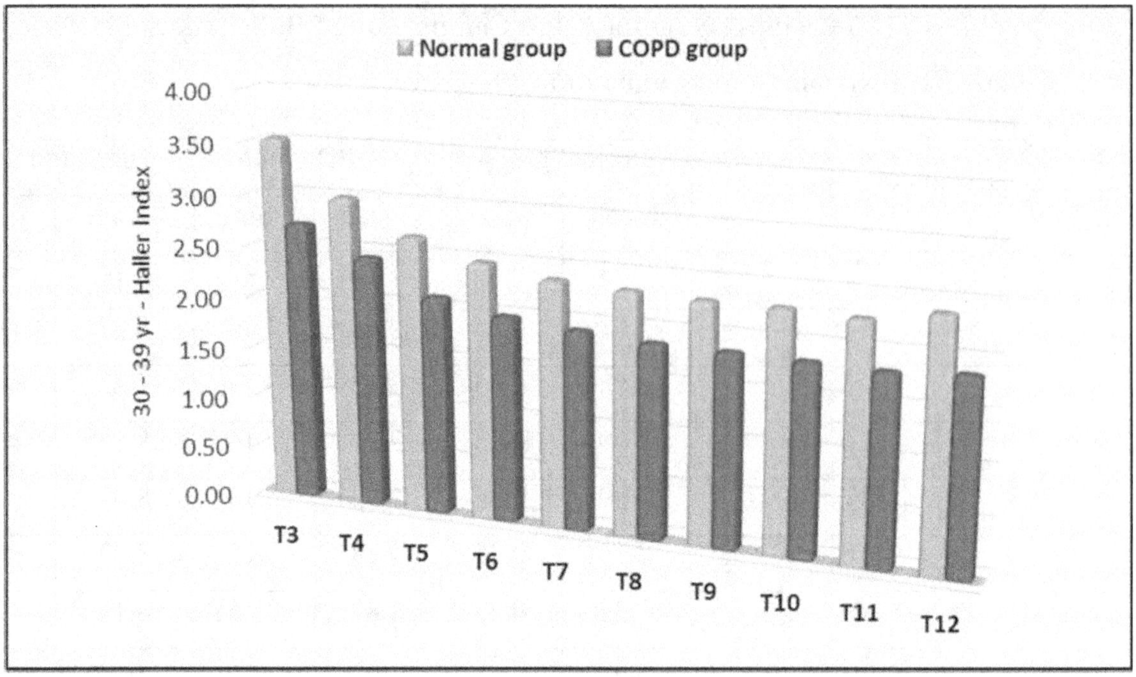

Graph No. - 1.26: Bar graph showing Haller index per vertebral level in age group 30-39 years among Normal and COPD groups.

- **Descriptive Summary and Comparison of Maximum Internal Transverse Diameter between Normal and COPD Groups among age group 40 – 49 years (Table-28)**

The descriptive summary and comparison of maximum internal transverse diameter between the normal and COPD groups among the age group 40-49 years using an unpaired t test revealed no significant association in any of the groups (Table 28).

Table – 28: Descriptive Summary and Comparison of Maximum Internal Transverse Diameter between Normal and COPD Groups among age group 40 – 49 years

40 - 49 yr -Maximum Internal Transverse Diameter	Normal group		COPD group		unpaired t test	
	Mean	SD	Mean	SD	t-value	p-value
T3	18.15	2.04	17.42	2.97	1.18	0.241
T4	20.39	1.54	19.56	2.45	1.67	0.099
T5	21.54	1.43	20.86	1.81	1.72	0.090
T6	22.26	1.58	21.59	1.73	1.66	0.102
T7	22.71	1.58	22.06	1.67	1.66	0.102
T8	22.99	1.77	22.56	1.79	0.99	0.326
T9	23.72	2.03	23.33	1.96	0.80	0.425
T10	24.17	1.98	24.08	1.97	0.18	0.857
T11	24.59	2.01	24.62	2.02	-0.06	0.951
T12	24.67	2.08	24.79	2.14	-0.22	0.824

Table no-28 demonstrated smallest transverse diameter at proximal vertebral level T3 and highest value at T12. Transverse diameters at each vertebral level were observed in an increasing trend from T3 to T12. Transverse vertebral diameters from T3 to T10 were lower in the COPD group among 49–49-year old age group. While a reverse pattern was observed from vertebral T11 to T12. The difference between the upper and lower thoracic segments

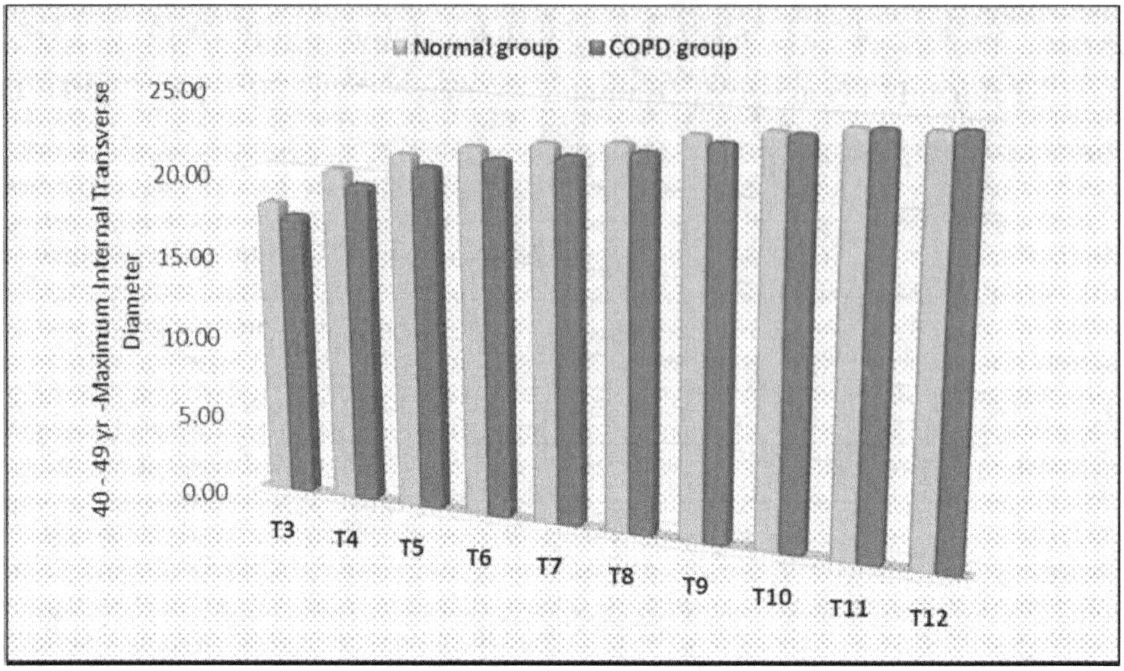

Graph No. - 1.27: Bar graph showing maximum internal transverse diameters per-vertebral level between Normal and COPD groups among age group 40-49 years.

- **Descriptive Summary and Comparison of Minimum Anteroposterior Diameter between Normal and COPD Groups among age group 40 – 49 years (Table-29)**

The findings, based on a descriptive summary and comparison of minimum anteroposterior diameter between normal and COPD groups in the age group 40-49 years, revealed a significant association in all groups with T3 (t=-3.85; p<0.001),T4 (t=-5.56; p<0.001),T5 (t=-6.72; p<0.001), T6(t=-7.06; p<0.001), T7(t=-7.06; p<0.001), T8(t=-6.58; p<0.001), T9(t=-6.37; p<0.001), T10(t=-6.19; p<0.001), T11(t=-5.22; p<0.001) and T12(t=-4.37; p<0.001) (Table 29).

Table – 29: Descriptive Summary and Comparison of Minimum Anteroposterior Diameter between Normal and COPD Groups among age group 40 – 49 year

40 - 49 yr -Minimum Anteroposterior Diameter	Normal group		COPD group		unpaired t test	
	Mean	SD	Mean	SD	t-value	p-value
T3	5.37	0.91	6.64	1.70	-3.85	**<0.001**
T4	6.97	1.13	8.78	1.54	-5.56	**<0.001**
T5	8.17	1.22	10.33	1.43	-6.72	**<0.001**
T6	9.06	1.40	11.51	1.46	-7.06	**<0.001**
T7	9.68	1.56	12.21	1.39	-7.06	**<0.001**
T8	9.92	1.76	12.56	1.53	-6.58	**<0.001**
T9	9.99	1.93	13.03	2.01	-6.37	**<0.001**
T10	10.47	1.98	13.63	2.23	-6.19	**<0.001**
T11	10.98	2.11	13.85	2.43	-5.22	**<0.001**
T12	10.85	2.39	13.48	2.57	-4.37	**<0.001**

Table No. 29 demonstrated that the minimum anteroposterior diameter was smallest at proximal vertebral T3 and largest at T11. Anteroposterior diameters were observed in an increasing pattern from vertebral level T3 to T11 and a slightly decreased value at T12 as compared to T11 in the age group 40–49 years among the normal and COPD groups. Anteroposterior diameter at the per-vertebral level was observed significantly higher in the COPD group as compared to the normal group.

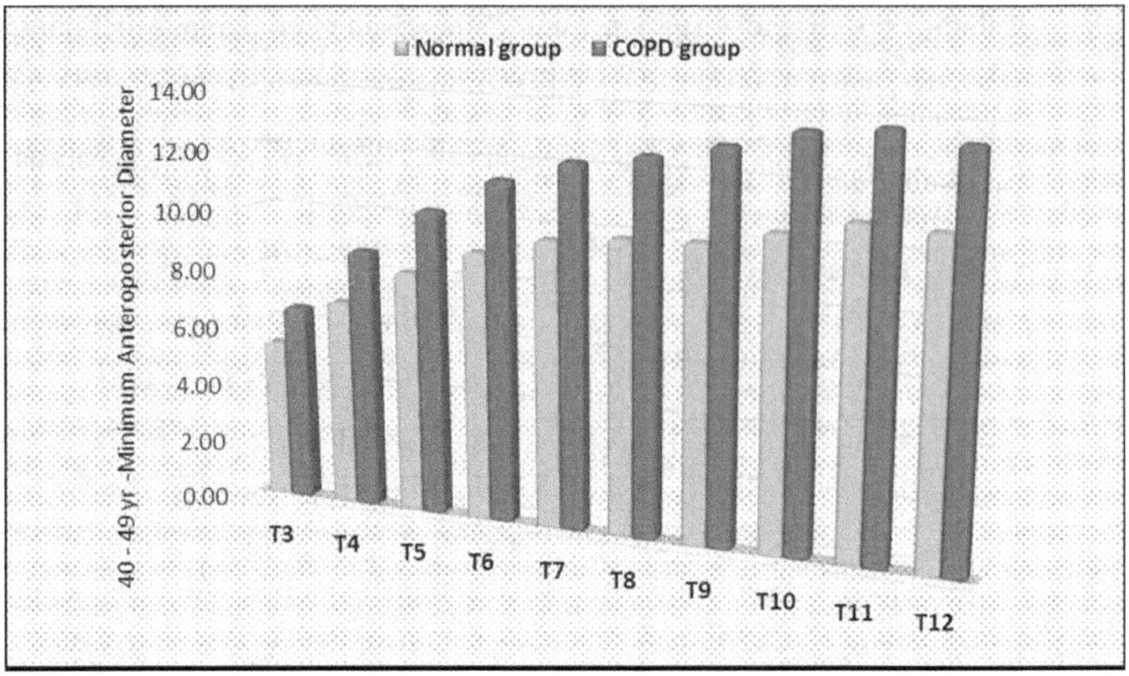

Graph No. - 1.28: Bar graph showing minimum anteroposterior diameter per-vertebral level between Normal and COPD groups among age group 40-49 years.

- **Descriptive Summary and Comparison of Haller Index between Normal and COPD Groups among age group 40 – 49 years (Table 30)**

The descriptive summary and comparison of the Haller index between normal and COPD groups among the age group 40-49 years using an unpaired t test revealed a significant association in all groups with T3 (t=5.35; p<0.001),T4 (t=7.01; p<0.001),T5 (t=6.49; p<0.001), T6(t=6.30; p<0.001), T7(t=-5.79; p<0.001), T8(t=3.45; p=0.001), T9(t=3.62; p=0.001), T10(t=4.13; p<0.001), T11(t=3.66; p=0.001) and T12(t=3.95; p<0.001) (Table 30).

Table – 30: Descriptive Summary and Comparison of Haller Index between Normal and COPD Groups among age group 40 – 49 year

40 - 49 yr -Haller Index	Normal group		COPD group		unpaired t test	
	Mean	SD	Mean	SD	t-value	p-value
T3	3.47	0.72	2.69	0.42	5.35	**<0.001**
T4	3.00	0.55	2.26	0.27	7.01	**<0.001**
T5	2.70	0.56	2.04	0.20	6.49	**<0.001**
T6	2.53	0.56	1.89	0.17	6.30	**<0.001**
T7	2.43	0.58	1.82	0.16	5.79	**<0.001**
T8	2.37	0.92	1.81	0.16	3.45	**0.001**
T9	2.55	1.15	1.82	0.20	3.62	**0.001**
T10	2.44	0.87	1.79	0.20	4.13	**<0.001**
T11	2.37	0.85	1.81	0.23	3.66	**0.001**
T12	2.40	0.71	1.88	0.25	3.95	**<0.001**

Table No. 30 demonstrated that the Haller index was highest at cranial vertebral T3 in the age group of 40–49 years among the normal and COPD groups. Haller indices per vertebral level were seen in a decreasing pattern from cranial vertebral level T3 to T8, and an increment at T9 and T10 as compared to T8, while T8 and T11 were found to have the same values among the normal and COPD groups in the age group 40–49 years. In the COPD group, the Haller indices per vertebral level were lower than in the normal group.

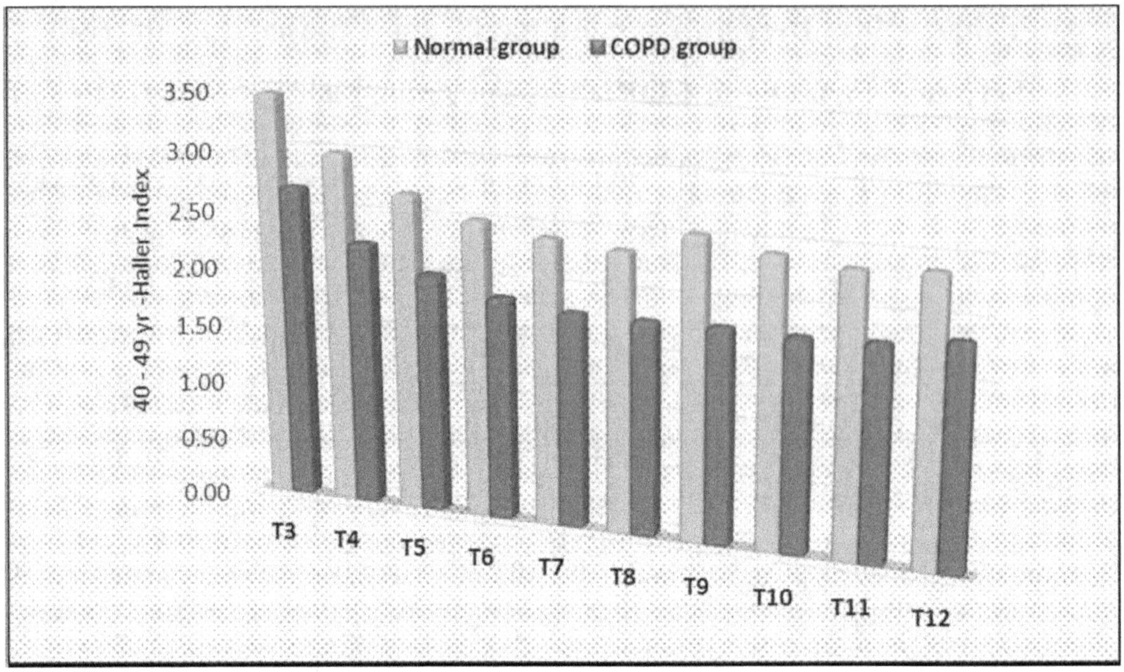

Graph No. - 1.29: Bar graph showing Haller index per vertebral level between Normal and COPD groups among age group 40-49 years.

❖ Descriptive Summary and Comparison of Maximum Internal Transverse Diameter between Normal and COPD Groups among age group 50 – 59 years (Table-31)

The findings, based on a descriptive summary and comparison of maximum internal transverse diameter between normal and COPD groups in the age group 50-59 years, revealed a significant association in all groups with T10 (t=-2.18; p=0.032), T11(t=-2.46; p=0.016) and T12(t=-2.64; p=0.010) (Table 31).

Table – 31: Descriptive Summary and Comparison of Maximum Internal Transverse Diameter between Normal and COPD Groups among age group 50 – 59 year

50 - 59 yr - Maximum Internal Transverse Diameter	Normal group		COPD group		unpaired t test	
	Mean	SD	Mean	SD	t-value	p-value
T3	17.57	2.16	17.67	2.02	-0.24	0.812
T4	19.69	1.89	19.87	1.69	-0.48	0.635
T5	20.77	1.66	21.00	1.61	-0.66	0.512
T6	21.49	1.68	21.78	1.62	-0.82	0.414
T7	22.00	1.73	22.32	1.63	-0.89	0.376
T8	22.37	1.83	22.80	1.74	-1.14	0.258
T9	22.83	1.94	23.51	1.82	-1.70	0.093
T10	23.45	2.08	24.42	2.07	-2.18	**0.032**
T11	23.99	2.21	25.09	2.00	-2.46	**0.016**
T12	24.27	2.32	25.48	2.00	-2.64	**0.010**

Table No. 31 demonstrated the smallest internal transverse diameter at proximal vertebral level T3 and the highest value at T12. Transverse diameters per vertebral level were observed in an increasing trend from T3 to T12. Transverse diameters at each vertebral level from T3 to T12 were lower in the normal group than in the 50-59 age group. Significant increments in the COPD group were observed at distal vertebral levels T10, T11, and T12.

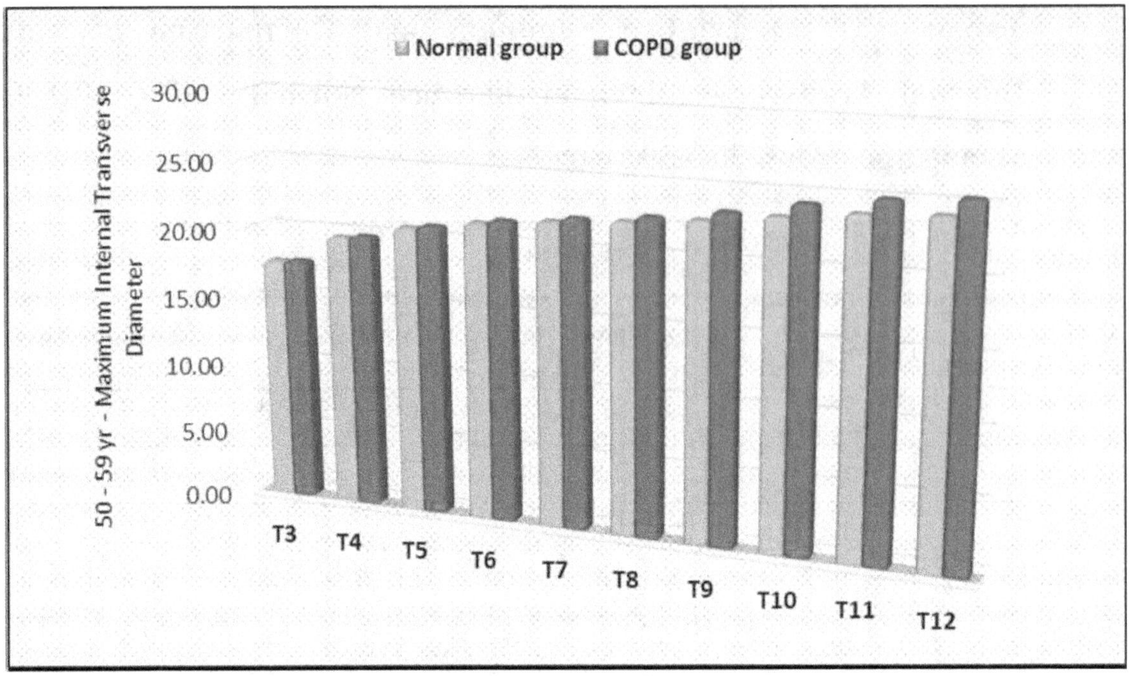

Graph No. - 1.30: Bar graph showing maximum internal transverse diameter per-vertebral level between Normal and COPD groups among age group 50-59 years.

- **Descriptive Summary and Comparison of Minimum Anteroposterior Diameter between Normal and COPD Groups among age group 50 – 59 years (Table-32)**

The findings, based on a descriptive summary and comparison of minimum anteroposterior diameter between normal and COPD groups in the age group 50-59 years, revealed a significant association in all groups with T3 ($t=-5.73$; $p<0.001$),T4 ($t=-6.80$; $p<0.001$),T5 ($t=-7.70$; $p<0.001$), T6($t=-8.63$; $p<0.001$), T7($t=-8.70$; $p<0.001$), T8($t=-7.80$; $p=0.001$), T9($t=-8.81$; $p=0.001$), T10($t=-8.96$; $p<0.001$), T11($t=-7.88$; $p=0.001$) and T12($t=-7.68$; $p<0.001$) (Table 32).

Table – 32: Descriptive Summary and Comparison of Minimum Anteroposterior Diameter between Normal and COPD Groups among age group 50 – 59 year

50 - 59 yr - Minimum Anteroposterior Diameter	Normal group		COPD group		unpaired t test	
	Mean	SD	Mean	SD	t-value	p-value
T3	5.09	0.98	6.66	1.48	-5.73	**<0.001**
T4	6.75	0.98	8.60	1.47	-6.80	**<0.001**
T5	8.00	1.03	10.10	1.44	-7.70	**<0.001**
T6	8.97	1.08	11.27	1.36	-8.63	**<0.001**
T7	9.58	1.20	12.01	1.39	-8.70	**<0.001**
T8	9.80	1.28	12.77	2.09	-7.80	**<0.001**
T9	9.96	1.46	12.99	1.72	-8.81	**<0.001**
T10	10.45	1.57	13.70	1.80	-8.96	**<0.001**
T11	10.84	1.79	13.97	1.90	-7.88	**<0.001**
T12	10.60	1.90	13.75	1.94	-7.68	**<0.001**

Table No. 32 demonstrated that the minimum anteroposterior diameter was smallest at proximal vertebral T3 and largest at T11. Anteroposterior diameters were observed in an increasing pattern from vertebral level T3 to T11 and a slightly decreased value at T12 as compared to T11 in the age group 50–59 years among the normal and COPD groups. Anteroposterior diameter at the per-vertebral level was significantly higher in the COPD group as compared to the normal group.

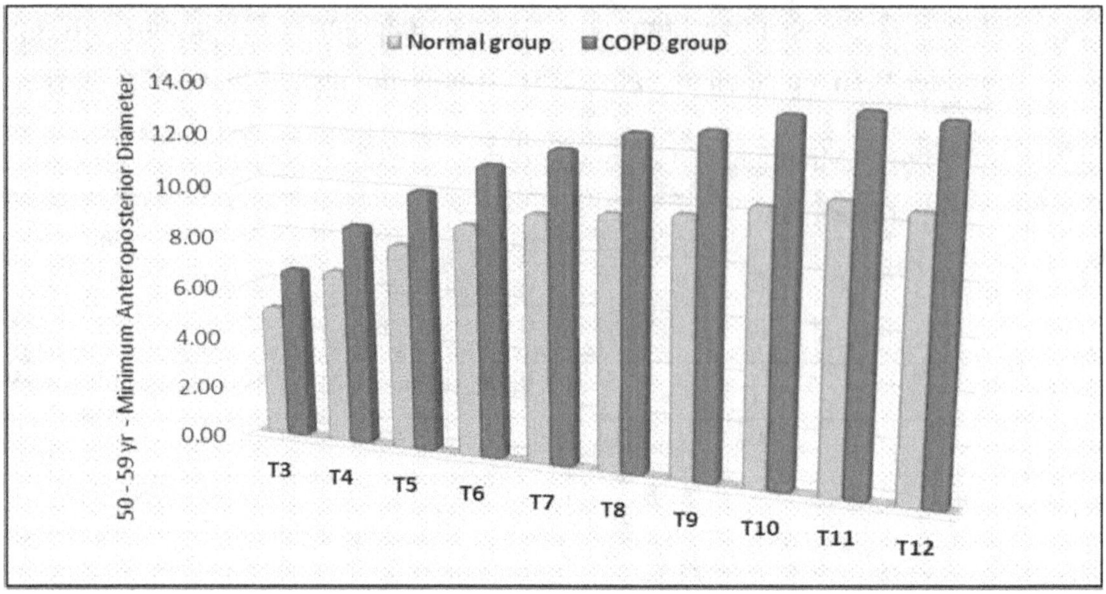

Graph No. - 1.31: Bar graph showing Anteroposterior Diameter between Normal and COPD Groups among age group 50 – 59 years.

- **Descriptive Summary and Comparison of Haller Index between Normal and COPD Groups among age group 50 – 59 years (Table-33)**

The descriptive summary and comparison of the Haller index between normal and COPD groups in the age group 50-59 years using an unpaired t test revealed a significant association in all groups with T3 ($t=7.36$; $p<0.001$),T4 ($t=8.40$; $p<0.001$),T5 ($t=8.95$; $p<0.001$), T6($t=9.58$; $p<0.001$), T7($t=9.12$; $p<0.001$), T8($t=9.27$; $p=0.001$), T9($t=-8.88$; $p=0.001$), T10($t=8.48$; $p<0.001$), T11($t=8.15$; $p=0.001$) and T12($t=8.08$; $p<0.001$) (Table 33).

Table – 33: Descriptive Summary and Comparison of Haller Index between Normal and COPD Groups among age group 50 – 59 years

50 - 59 yr - Haller Index	Normal group		COPD group		unpaired t test	
	Mean	SD	Mean	SD	t-value	p-value
T3	3.53	0.52	2.74	0.48	7.36	**<0.001**
T4	2.96	0.35	2.36	0.32	8.40	**<0.001**
T5	2.63	0.30	2.11	0.25	8.95	**<0.001**
T6	2.42	0.27	1.95	0.19	9.58	**<0.001**
T7	2.32	0.28	1.87	0.18	9.12	**<0.001**
T8	2.31	0.26	1.82	0.24	9.27	**<0.001**
T9	2.33	0.32	1.83	0.21	8.88	**<0.001**
T10	2.29	0.34	1.80	0.20	8.48	**<0.001**
T11	2.25	0.31	1.82	0.20	8.15	**<0.001**
T12	2.34	0.32	1.88	0.22	8.08	**<0.001**

Table No. 33 demonstrated that the Haller index was highest at cranial vertebral T3 in the age group 50–59 years among the normal and COPD groups. The Haller indices per vertebral level decreased from cranial vertebral level T3 to T11, with the exception of T9, which was slightly higher than T8 and T10, and an increased value was observed at T12 compared to T11 in the Normal group. In the COPD group, the Haller index decreased from T3 to T8, then increased insignificantly at T9, followed by another decrease at T10 and T11. In the COPD group, incremental change at T12 was also observed as compared to T11. The COPD group had significantly lower Haller indices per vertebral than the normal group.

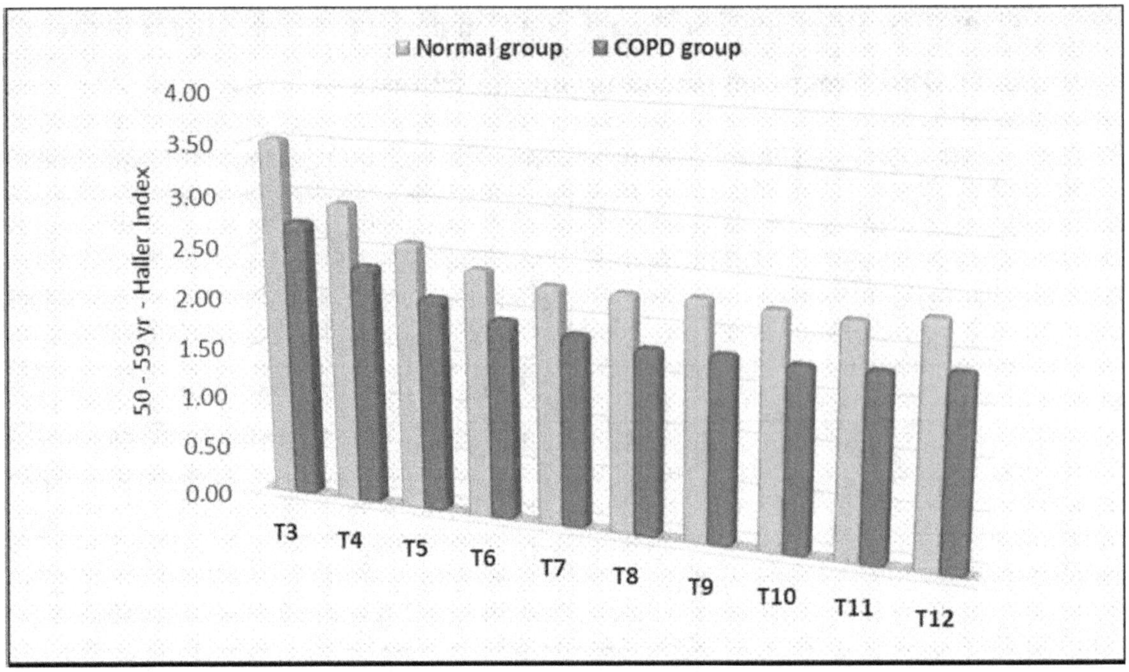

Graph No. - 1.32: Bar graph showing Haller Index per-vertebral level between Normal and COPD Groups among age group 50 – 59 years.

- ### Descriptive Summary and Comparison of Maximum Internal Transverse Diameter between Normal and COPD Groups among age group 60 – 70 years (Table-34)

The findings, based on a descriptive summary and comparison of maximum internal transverse diameter between normal and COPD groups in the age group 60-70 years, revealed a significant association in all groups with T3 (t=2.03; p=0.045),T4 (t=2.47; p=0.015),T5 (t=3.15; p=0.002), T6(t=2.86; p=0.005), T7(t=2.76; p=0.007) and T8(t=2.22; p=0.029) (Table 34).

Table – 34: Descriptive Summary and Comparison of Maximum Internal Transverse Diameter between Normal and COPD Groups among age group 60 – 70 year

60 - 70 yr - Maximum Internal Transverse Diameter	Normal group		COPD group		unpaired t test	
	Mean	SD	Mean	SD	t-value	p-value
T3	18.61	1.89	17.69	2.24	2.03	**0.045**
T4	20.75	1.68	19.78	1.91	2.47	**0.015**
T5	22.11	1.77	20.93	1.76	3.15	**0.002**
T6	22.65	1.67	21.60	1.76	2.86	**0.005**
T7	23.04	1.64	22.00	1.83	2.76	**0.007**
T8	23.31	1.70	22.45	1.89	2.22	**0.029**
T9	23.69	1.81	23.19	2.08	1.17	0.245
T10	24.35	2.04	24.12	2.22	0.51	0.610
T11	24.91	2.19	24.83	2.23	0.18	0.856
T12	25.20	2.36	25.08	2.87	0.21	0.831

Table No. 34 demonstrated the smallest internal transverse diameter at proximal vertebral level T3 and the highest value at T12. Transverse vertebral diameters were found to increase in both groups from T3 to T12. Transverse vertebral diameters from T3 to T8 were significantly lower, while vertebral levels T9 to T12 were insignificantly lower in the COPD group of 50-59-years old. A difference between the upper T3 to T8 and lower segments T9 to T12 was observed.

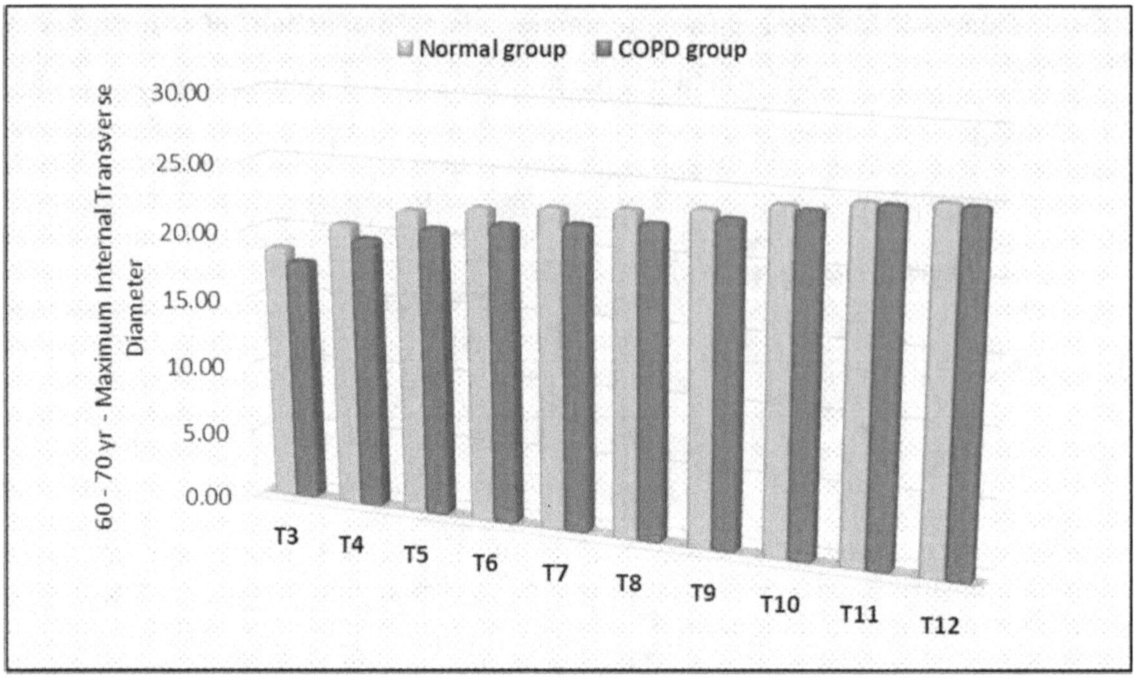

Graph No. - 1.33: Bar graph showing maximum internal transverse diameter per-vertebral level between Normal and COPD groups among age group 60-70 years.

- **Descriptive Summary and Comparison of Minimum Anteroposterior Diameter between Normal and COPD Groups among age group 60 – 70 years (Table-35)**

The findings, based on a descriptive summary and comparison of minimum anteroposterior diameter between normal and COPD groups in the age group 60-70 years, revealed a significant association in all groups with T3 (t=-4.67; p<0.001),T4 (t=-5.88; p<0.001),T5 (t=-7.48; p<0.001), T6(t=-8.68; p<0.001), T7(t=-8.52; p<0.001), T8(t=-8.66; p<0.001), T9(t=-8.36; p<0.001), T10(t=-8.04; p<0.001), T11(t=-7.14; p<0.001) and T12(t=-5.48; p<0.001) (Table 35).

Table – 35: Descriptive Summary and Comparison of Minimum Anteroposterior Diameter between Normal and COPD Groups among age group 60 – 70 years

60 - 70 yr - Minimum Anteroposterior Diameter	Normal group		COPD group		unpaired t test	
	Mean	SD	Mean	SD	t-value	p-value
T3	5.69	0.86	6.87	1.31	-4.67	**<0.001**
T4	7.27	0.96	8.92	1.47	-5.88	**<0.001**
T5	8.43	0.96	10.50	1.44	-7.48	**<0.001**
T6	9.40	0.93	11.62	1.32	-8.68	**<0.001**
T7	10.00	1.06	12.35	1.40	-8.52	**<0.001**
T8	10.04	1.33	12.67	1.48	-8.66	**<0.001**
T9	10.23	1.20	13.12	1.79	-8.36	**<0.001**
T10	10.52	1.47	13.55	1.91	-8.04	**<0.001**
T11	10.87	1.79	13.87	2.05	-7.18	**<0.001**
T12	10.57	2.16	13.39	2.54	-5.48	**<0.001**

Table No. 35 demonstrated that the minimum anteroposterior diameter was smallest at proximal vertebral T3 and largest at T11 in both groups. Anteroposterior diameters increased from vertebral level T3 to T11, with a slightly lower value at T12 compared to T11 in the age group 60-70 years in the Normal and COPD groups. Anteroposterior diameter at the per-vertebral level was significantly higher in the COPD group as compared to the normal group.

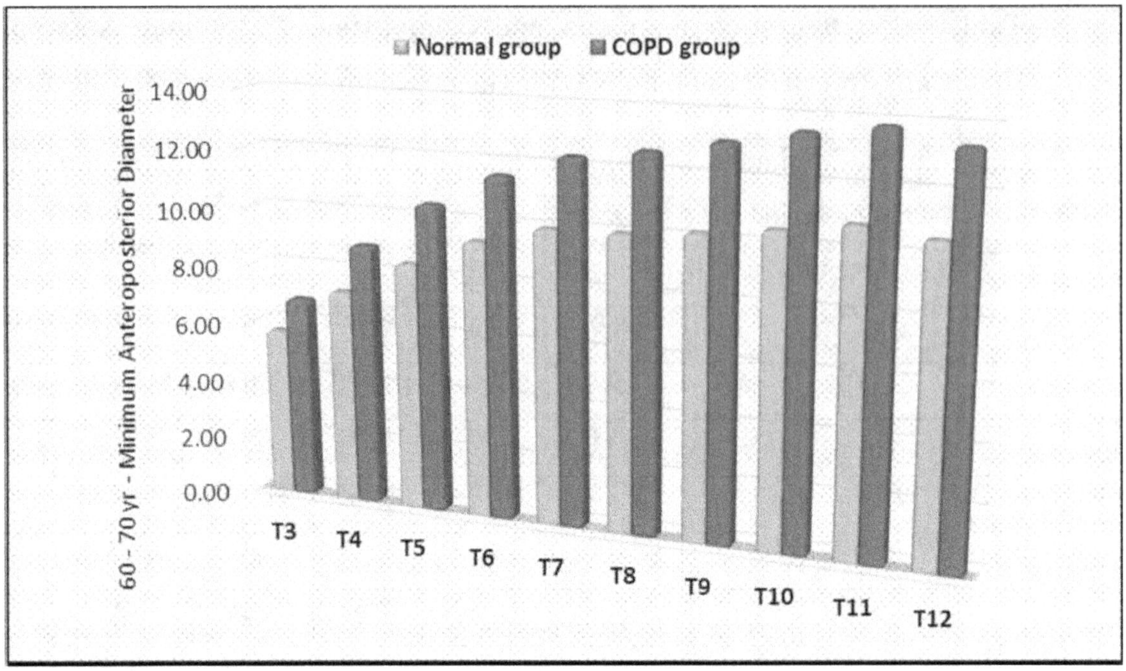

Graph No. - 1.34: Bar graph showing minimum anteroposterior diameter per-vertebral level between Normal and COPD Groups among age group 60 – 70 years.

- ❖ **Descriptive Summary and Comparison of Haller Index between Normal and COPD Groups among age group 60 – 70 years (Table-36)**

A significant association was found in all groups with T3 in the results-based summary and comparison of the Haller index between normal and COPD groups among the age group 60-70 years using an unpaired t test (t=7.47; p<0.001),T4 (t=9.34; p<0.001),T5 (t=10.89; p<0.001), T6(t=12.14; p<0.001), T7(t=11.56; p<0.001), T8(t=10.32; p<0.001), T9(t=10.66; p<0.001), T10(t=10.77; p<0.001), T11(t=9.66; p<0.001) and T12(t=7.09; p<0.001) (Table 36).

Table – 36: Descriptive Summary and Comparison of Haller Index between Normal and COPD Groups among age group 60 – 70 year

60 - 70 yr - Haller Index	Normal group		COPD group		unpaired t test	
	Mean	SD	Mean	SD	t-value	p-value
T3	3.31	0.44	2.64	0.42	7.47	**<0.001**
T4	2.89	0.32	2.26	0.32	9.34	**<0.001**
T5	2.65	0.31	2.02	0.25	10.89	**<0.001**
T6	2.43	0.23	1.88	0.20	12.14	**<0.001**
T7	2.32	0.26	1.80	0.19	11.56	**<0.001**
T8	2.36	0.36	1.79	0.20	10.32	**<0.001**
T9	2.34	0.30	1.79	0.21	10.66	**<0.001**
T10	2.35	0.31	1.80	0.20	10.77	**<0.001**
T11	2.34	0.35	1.81	0.19	9.66	**<0.001**
T12	2.45	0.40	1.92	0.33	7.09	**<0.001**

Table No. 36 demonstrated that the Haller index was largest at proximal vertebral T3 and smallest at T11, with an increment at T12 as compared to T11 in the normal group. Haller indices were observed in a decreasing pattern from vertebral level T3 to T11 except for T10 and a slightly increased value at T12 as compared to T11 in the age group 50–59 years among the normal group. In the COPD group, Haller indices were observed in a decreasing pattern from T3 to T9, with insignificant increments at T10 and T11, and finally an increase at T12 as compared to T11. The COPD group had significantly lower Haller indices per vertebral level than the normal group.

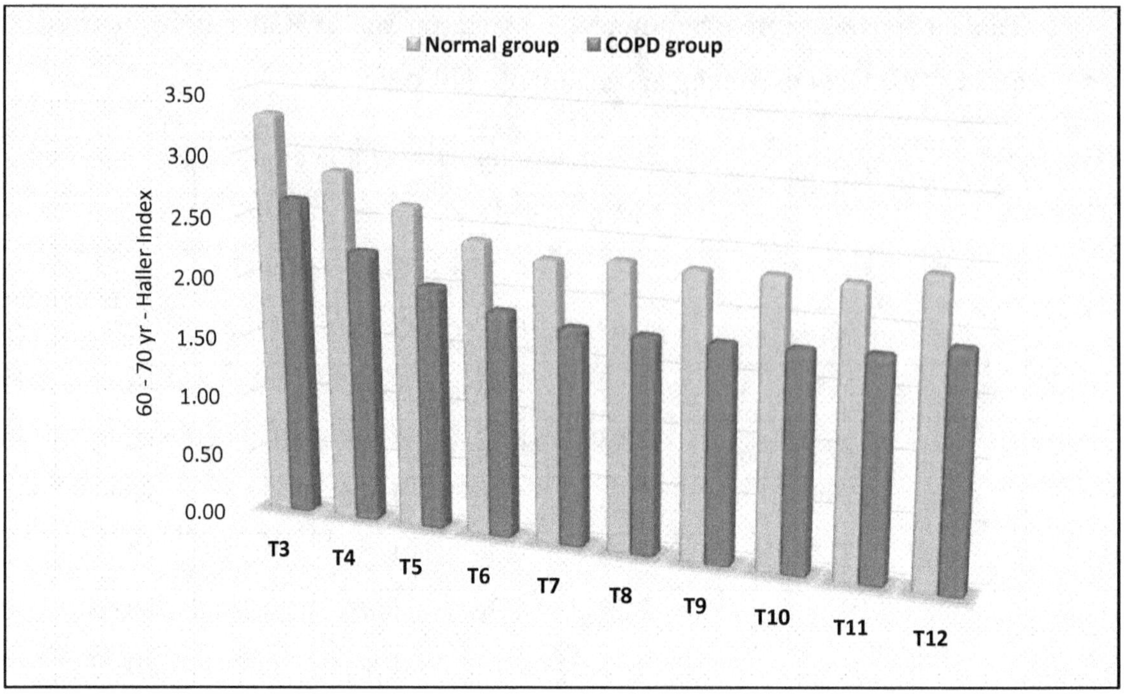

Graph No. - 1.35: Bar graph showing Haller index per-vertebral level between Normal and COPD Groups among age group 60 – 70 years.

❖ Descriptive Summary and Comparison of Average Maximum Internal Transverse Diameter between Normal and COPD Groups and between Males & Females (Table-37)

The results showed that no significant difference in average maximum internal transverse diameter was found between the normal and COPD groups among males ($p=0.485$) and females ($p=0.083$).

Among the normal group, a significant difference was found between males and females ($p<0.001$). The average maximum internal transverse diameter of males was higher than that of females.

Among the COPD group, a significant difference was found between males and females ($p<0.001$). The average maximum internal transverse diameter of males was higher than that of females.

Table – 37: Descriptive Summary and Comparison of Average Maximum Internal Transverse Diameter between Normal and COPD Groups and between Males & Females

Maximum Internal Transverse Diameter	Normal group				COPD group				unpaired t test	
	Mean	SD	Min	Max	Mean	SD	Min	Max	t-value	p-value
Male	23.56	1.52	19.11	27.11	23.41	1.36	19.55	26.71	0.70	0.485
Female	21.27	1.23	18.31	24.83	20.93	1.46	14.78	26.81	1.74	0.083
Male vs Female	**t=11.47, p<0.001**				**t=12.18, p<0.001**					

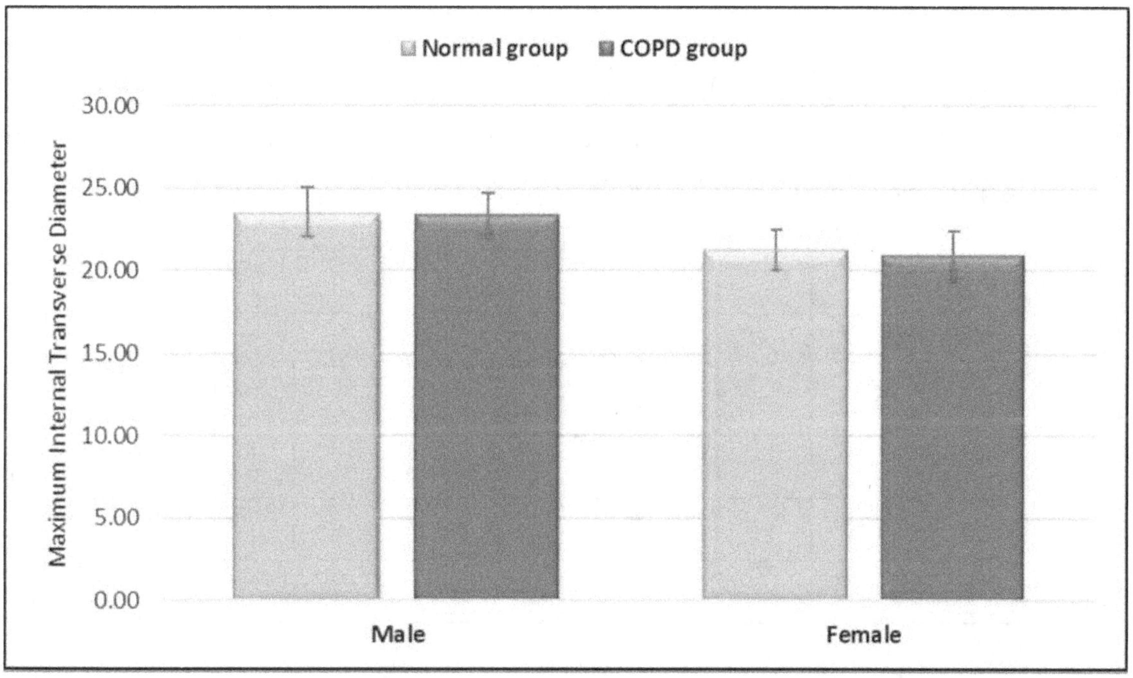

Graph No. - 1.36: Bargraph showing Average Maximum Internal Transverse Diameter between Normal and COPD Groups and between Males & Females

❖ Descriptive Summary and Comparison of Average Minimum Anteroposterior Diameter between Normal and COPD Groups and between Males & Females (Table-38)

The results showed that a significant difference in average minimum anteroposterior diameter was found between the normal and COPD groups among males ($p<0.001$) and females ($p<0.001$). It was higher among COPD cases compared to normal cases.

Among the normal group, a significant difference was found between males and females ($p<0.001$). The average minimum anteroposterior diameter of males was greater than that of females.

Among the COPD group, a significant difference was found between males and females ($p<0.001$). The average minimum anteroposterior diameter of males was greater than that of females.

Table – 38: Descriptive Summary and Comparison of Average Minimum anteroposterior Diameter between Normal and COPD Groups and between Males & Females

Minimum Anteroposterior Diameter	Normal group				COPD group				unpaired t test	
	Mean	SD	Min	Max	Mean	SD	Min	Max	t-value	p-value
Male	9.63	1.08	6.98	11.76	12.18	1.37	9.02	16.53	-14.28	**<0.001**
Female	8.57	1.06	3.17	10.36	10.90	1.23	8.11	15.10	-14.03	**<0.001**
Male vs Female	**t=6.86, p<0.001**				**t=6.81, p<0.001**					

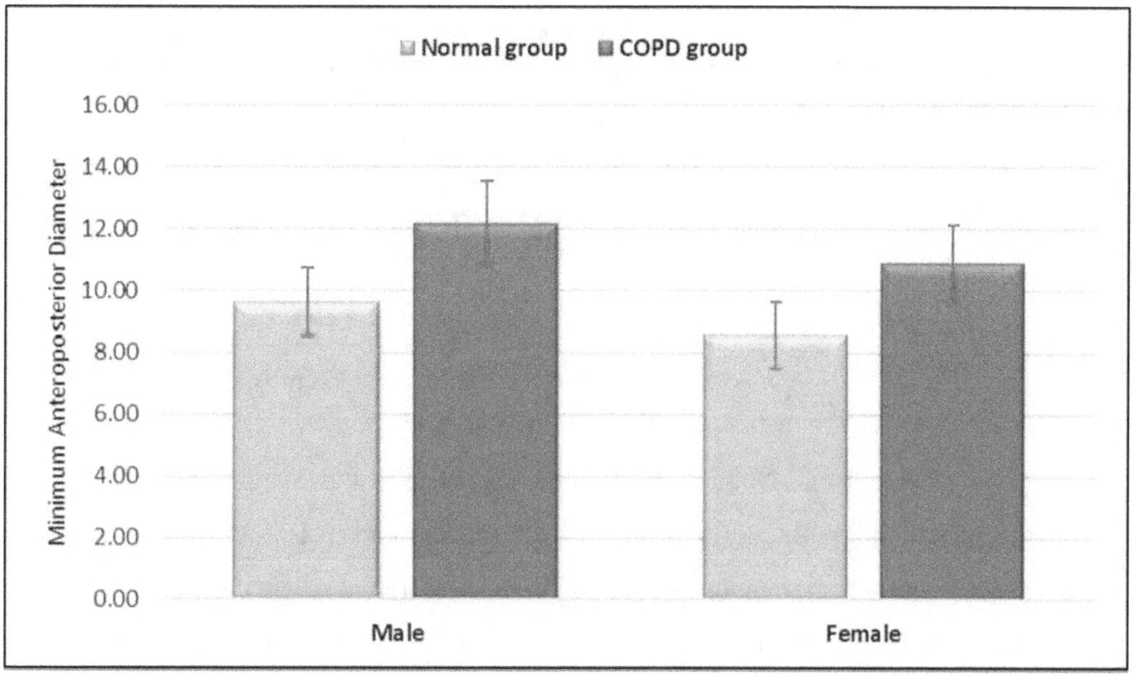

Graph No.- 1.37: Bar graph showing Average Minimum Anteroposterior Diameter between Normal and COPD Groups and between Males & Females

- **Descriptive Summary and Comparison of Average Haller Index between Normal and COPD Groups and between Males & Females (Table-39)**

The results showed that a significant difference in average Haller index was found between the normal and COPD groups among males ($p<0.001$) and females ($p<0.001$). It was smaller among COPD cases compared to normal cases.

Among the normal group, no significant difference was found between males and females ($p=0.364$).

Among the COPD group, no significant difference was found between males and females ($p=0.739$).

Table – 39: Descriptive Summary and Comparison of Average Haller index with minimum and maximum range between Normal and COPD Groups and between Males & Females

Haller Index	Normal group				COPD group				unpaired t test	
	Mean	SD	Min	Max	Mean	SD	Min	Max	t-value	p-value
Male	2.55	0.26	2.26	3.30	2.00	0.18	1.49	2.26	16.97	**<0.001**
Female	2.60	0.46	2.26	3.43	1.99	0.19	1.43	2.71	11.96	**<0.001**
Male vs Female	t=0.91, p=0.364				t=0.33, p=0.739					

According to Table No. 39, the overall mean Haller index in male controls was 2.55. In the current study, the normal range of the Haller index was 2.26 to 3.30. For females, the overall mean Haller index was 2.60 for normal females. In the normal group, females' Haller index ranges between 2.26 and 3.43. The overall mean Haller index in the COPD male group was 2.00. It varied between a minimum of 1.49 and a maximum of 2.26. The overall Haller in the COPD female group is 1.99. It ranged from a low of 1.43 to a high of 2.71.

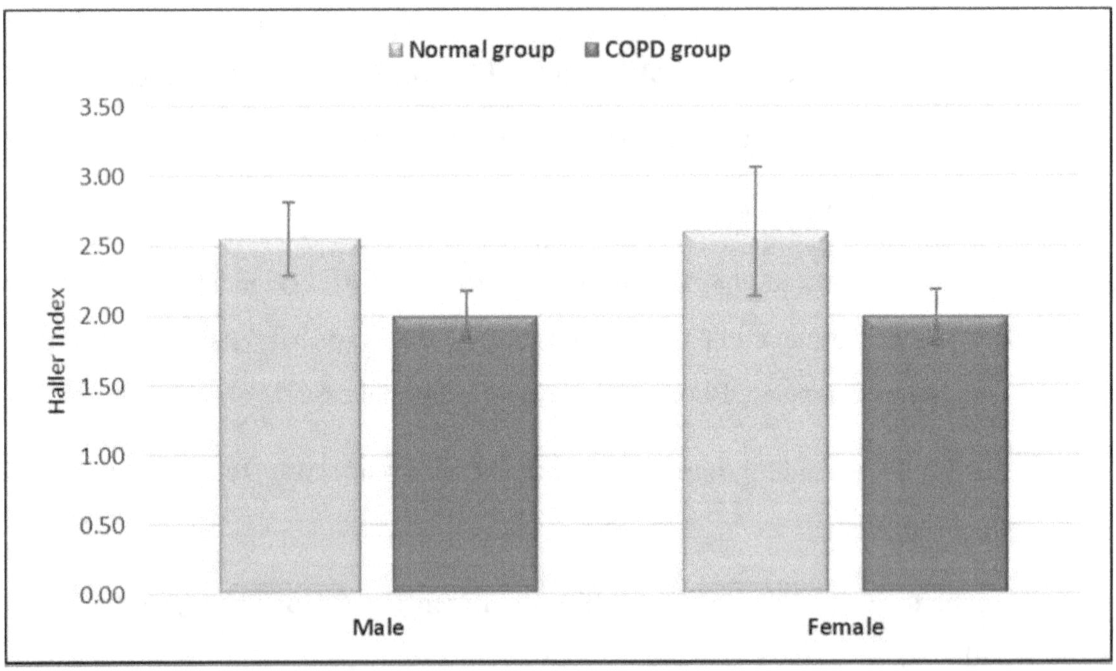

Graph No. - 1.38: Bar graph showing Average Minimum Anteroposterior Diameter between Normal and COPD Groups and between Males & Females

DISCUSSION

In humans, the ribcage is a complex three-dimensional osteochondral armature. Its structure represents vital intrathoracic organs such as the lungs and heart. The chest protects vital organs while providing attachment points for extrinsic and intrinsic muscles. These muscles directly or indirectly support the respiratory process necessary for survival. The rostral end of the ribcage is created by the higher ribs that form the lower rim of the thoracic entrance contributing to the downward slope of the ribs (J E Archer et. al)[77]. It can be argued that it is not a true chest (J E Archer et. al)[77]. The thorax, thin on top and continuing from the base of the neck, resembles a truncated cone. The distal end of the thorax is wider but forms a thoracic outlet separated from the abdominal cavity by the diaphragm, the tendinous muscle layer. The rib cage is flat dorsoventrally and convex laterally.

The results showed that the mean transverse and anteroposterior dimensions were the shortest at T3. Moreover, there was a consistent increment at each vertebral level up to T12 for transverse diameter. Both diameters were the largest at caudal vertebral level but there was slightly reduced value at T12 for anteroposterior diameter as compared to T11. Similar trends were observed for maximum and minimum variable values with both measured transverse and anteroposterior diameters, when mean values were omitted. This fact strongly favors a tapered top and a wide bottom near the exit. However, the lateral diameter is small in infancy but begins to increase as a person learns to walk (Henry Gray-545-46)[192]. Cross-sectional area peaks in adulthood (Henry Gray-545-46)[192]. The maximum size of the transverse diameter, as well as the convexity and curvature of the ribs, make the chest laterally convex (Henry Gray-537)[1]. Because the anteroposterior diameter was more than three times smaller than the lateral diameter at T3. Measured anteroposterior diameters were found to be nearly twofold smaller at each vertebral level subsequently. This fact indicates a dorsoventral flat shape of the chest.

The sternum forms the anterior margin of the thoracic entrance at vertebral level T3. The muscles of the anterior abdominal wall form the anterior border of the ribcage below the distal end of the sternum. On examining the 3-D reconstructed rib cage from the outside between T1 and T12, the anterior midline border is formed by the cranial end, body and caudal end of the sternum in front and behind thoracic segment of the vertebral column. It showed that the posterior boundary formed by thoracic segment of vertebral column in midline was appeared longer than the front border formed by sternum in front in the mid line. Actually, it was much longer (see Figure 2.03) too.

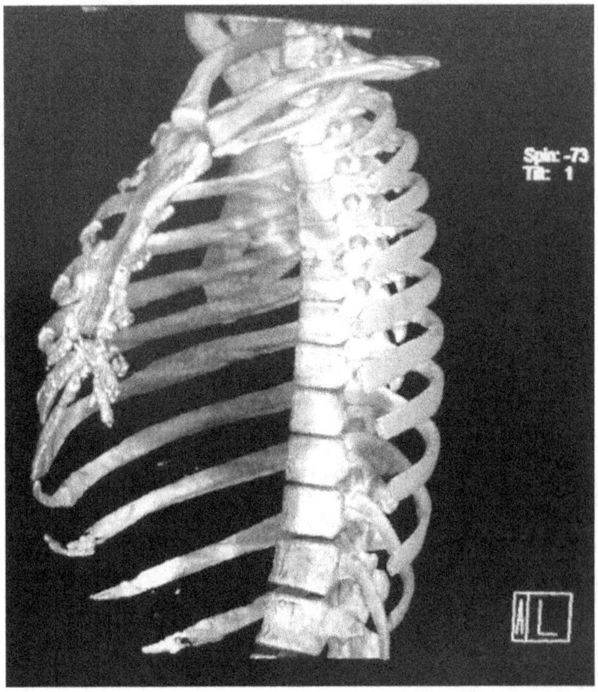

Figure No: 3.20

3D computed tomography reconstructed thoracic cage image of a 40 years' female with cut ribs on the left side showing anterior and posterior borders of the thoracic cage.

In some cases, we performed three-dimensional computed tomography (CT) reconstructions of several chest cage images. We found an interesting case where the anterior end of the right 4th rib was bifurcated (ref figure no: 3.20). Bifurcated ribs are a rare congenital condition in which the tip of the ribs is bifurcated. Dichotomy of the rib can be unilateral or bilateral in nature. In populations, the prevalence of dichotomy ranges from 0.15% to 3.4% (E. Kupeli et. al.)[193]. Ribs account for up to 20% of congenital rib abnormalities (E. Kupeli et. al.)[193]. Gorin-Goltz syndrome can be distinguished by bipartite ribs (J. A. Geel)[194]. Physicians, Surgeons, and Radiologists are all concerned about rib bisection.

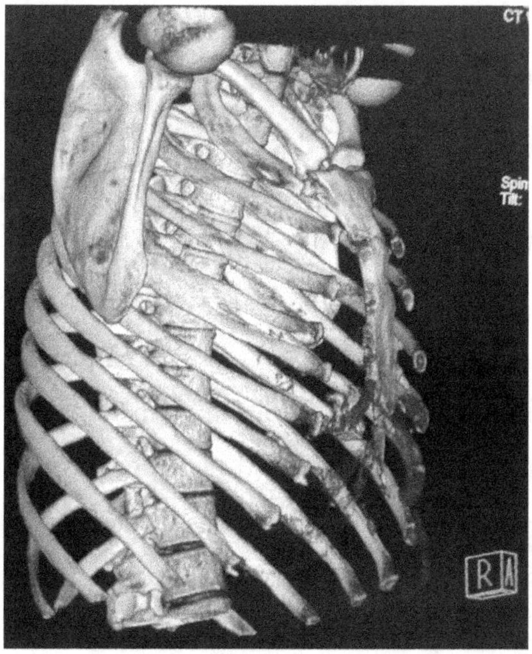

Figure No: 3.21

Three-dimensional computed tomography reconstructed the thoracic cage image of a 36 years old female showing a bifurcated anterior end of the 4th right rib reported in the Normal group.

When viewing a chest Computed Tomography image in the axial plane, the chest begins at the T3 vertebra level below the thoracic inlet. On axial images, there was no consistent discernible bone showing the anterior portion above T3. The sternum forms the anterior border of the chest from vertebral level T3 to T10. The rectus abdominis muscle forms the anterior border of the rib cage below the distal end of the sternum (T10), and the linea-alba that forms the midline is formed by the oblique fibers of the aponeurosis of the anterior abdominal flat muscle wall. Haller index was created and developed by Haller et al.[78]. This index was created by dividing the largest thoracic cross-sectional diameter assessed on axial thoracic Computed Tomography images by the smallest anteroposterior diameter. This metric was used to assess the funnel chest. This index was calculated and measured at the single vertebra level for each subject. As a result, the spinal level was not fixed and referenced. Haller et. al.[78] estimated the mean index from a small sample size of 2.56, with no thoracic cage deformities in their original study. In the current study we reported overall mean Haller 2.55 for male and 2.60 for female among normal group. This finding was consistent with Haller et. al.[78].

Daunt et al.[79] when then analyzed numerous chests Computed Tomography images of children from 0 to 18 years without congenital chest deformities. They used the height of the thoracic cavity between vertebrae and sternum to determine the index which gave the highest values for the Haller index, but this was not always the same at each vertebral level. Normal values were shown to be lower than in the previous studies (2.56). This study (Daunt et al.) found that the Haller Index was age dependent. As a normal person develops, the Haller Index changes. JE Archer et al.[77] for their study, used a large number of Computed Tomography images of the peadiatric population from a normal person aged between 8 to 18 years. Multiple ten vertebral levels of the Haller index were established and included in their study (T3 to T12). It was emphasized that a single average Haller index value was insufficient. The Haller index behaves differently at each vertebra level. In the present study it was discovered that the maximum value (mean) of the Haller index 3.52 in male and 3.49 in females were detected at T3. Then a decreasing trend to T8 and T9, followed by higher values at T10, and T11 in comparison to T9, and at T12 finally increased value as compared to T11 in both sexes. In present study we reported slightly higher values of Haller index in females as compared to normal males which was statistically insignificant Except few vertebral levels. Result of the current study was similar as reported by JE Archer et. al.[77]. The Haller index was found increased with increasing age up to the age group of 18 to 49 years, especially in the cranial and less so in the caudal thorax, in both sexes among the normal group. Among the normal group between 60 to 70 years Haller indices per vertebral were observed lower as compared to other age groups.

Examining the effect of age, we found that the mean Haller index was increased by an average of 88.2% from adulthood (from age group 18-29 years) to middle age (40 to 49 years) at the vertebral column T3 level. This suggests that the anteroposterior diameter at the T3 level decreased with age. Age was positively correlated with the Haller index in middle-aged people. While Haller indices were observed to decrease in the old age (50 to 59 years and 60 to 70 years age group) group as compared to the age group 18 to 49 years among normal. While there was no regular pattern, the Haller Index at the T4 level was greater in 94% instances as compared to early adulthood. The same applies to the T5 level. The Haller index was greater at the T6 level in 76.47% instances among normal patients. In 70.58% of instances, the Haller Index was greater at T7 and T8 levels. In 52.94% of cases, the Haller index was higher in middle aged people (30-49 years) as compared with the 18 to 29 years old age group. Haller Index was higher in middle aged people in 52.82% of patients at T10 vertebral level. Another decreasing trend was observed at the T11 and T12 levels as the age advanced.

The current study was unique in that the Haller index determined at ten multiple (T3 to T12) vertebral levels on CT chest images has rarely been studied in the North Indian population. Second, this study was conducted to determine the Haller Index in adult to middle-aged (18 to 49 years old) and from middle-aged to old (50 to 59 and 60 to 70 years) age groups in north Indian men and women. The Haller index was highest at T3, the first detectable vertebral level, in both males and females. This observation was consistent with that of JE Archer et al. The Haller index drops from T3 to T7. Further rises slightly at T8 and T9, drops again at T10 and T11, and rises again at T12 as compared to T11 in normal females. On an average decreasing pattern from T3 to T11 because particular alteration at each level was very minor. Vertebral level T7 and T10 had same values among normal females. Vertebral level T11 had lowest values of Haller index for normal women. This finding minutely differs from that reported by JE Archer et al.[77]. It was also clear that women have a higher Haller index per vertebral level than men. This finding was same as reported by JE Archer et al.[77]. An interesting aspect was that the difference between mean Haller indices for adult-to-middle-aged men and women was not statistically significant at each vertebral level. This can be said that after attaining the full growth there was no significant change in the curvature of vertebral column in early adulthood to middle aged men and woman. In the past study conducted by Daunt et. al. and JE Archer et. al.[77]. included child age group (0 to 18 and 8 to 18 years' age group). In their study the difference between male female was statistically noticeable because of the growth rate, the curvature of the spine changed with age of an individual during growing phase. In young adults and middle-aged people, the Haller index refers to the curvature of the vertebral column and varies with the thickness (height) of the intervertebral discs.

The results of the present study were significant as they were the first to describe the effect of age on chest shape in adult to old-aged North Indians. This document of normative data can be used for further research. The clinical importance is related to the surgical treatment and management of middle-aged to old-aged patients with funnel chest, pigeon chest, and undiagnosed scoliosis. Today, funnel deformity (pectus excavatum) of breast bone (sternum) is treated by minimally invasive surgical method with 'Nuss bars' in teenagers and young adults without exception reported by Nuss D et. al.[157]. The pigeon chest deformity is corrected by a surgical procedure called the Ravitch procedure based unique work by Pilegaard H and RAVITCH MM et. al.[195,196]. Scoliosis is an abnormal sideways curvature of the spine with rotation of the thoracic cage and development of a rib hump (Kálmán A)[197]. The Ravitch procedure is also the current surgical technique to correct the curvature of the spine and rotation of the thoracic cage (Hong J-Y)[198]. Therefore, it is critically important for corrective and reconstructive surgeons to know what is' normal' in surgical technique. Where the surgeon intends to return the patients to' normal'.

The Haller index measurement as a quantification was not without problems. There was evidence that there are differences between the Haller exponents. Many factors affect the quality of the obtained chest computed tomography (CT) image. The most fundamental factors include the Number of breathing stages of inspiration and expiration (Albertal M, Birkemeier K)[199, 200] and the associated amount of air volume inhaled and exhaled. In routine practice, computed tomography (CT) images were always acquired under conditions of maximal inspiration and breath-holding, so there was no reason to speculate about the scope of this study. This study measures the mean Haller index by vertebral column levels from T3 to T12 in healthy young adults, middle-aged, and old-aged normal subjects. This data pool, which includes men and women between the ages of 18-29, 30-39, 40-49, 50-59- and 60-70-years age group among control, provides a standard data record of undeformed chests. This information is useful for decision-making regarding thoracic dimensional distortion in patients with chronic obstructive pulmonary disease. Most of the COPD patients are hale in their 50's and their 70s. In previous studies mean age of COPD patients was reported 60 years. Our results also showed highest number of COPD patients were sixty-eight (68), constituted 35.4% of the total COPD participants lie in age group 60-70 years (refer table no-1). It would be of great benefit because in this era the scientists are attempted lung transplantation in severe chronic obstructive pulmonary disease (COPD) patients (Kasper DL et. al.)[201].

The present study reported an unusual case of pectus excavatum in a 45 years old female. She had consulted the Department of Pulmonary Medicine regarding recurrent attacks of breathlessness. She also felt breathless on routine walks and unable to climb stairs for the last 10 to 12 years. With breathlessness, she also felt palpitation. On physical examination her chest contour was abnormal, and the sternum bone was depressed deeply backward. She was advised CT thorax. Computed tomography findings for lung parenchyma were normal. On three-dimensional CT reconstruction of the thoracic cage, it revealed depressed breast bone (sternum), and anterior ends of the costal cartilages and ribs were projected more forwards. When Haller indices were measured in axial plane images from T3 to T12, it found always higher than 6.0. The anteroposterior diameter was found shorter than Normal and COPD groups among males and females. Though she was excluded from the study groups.

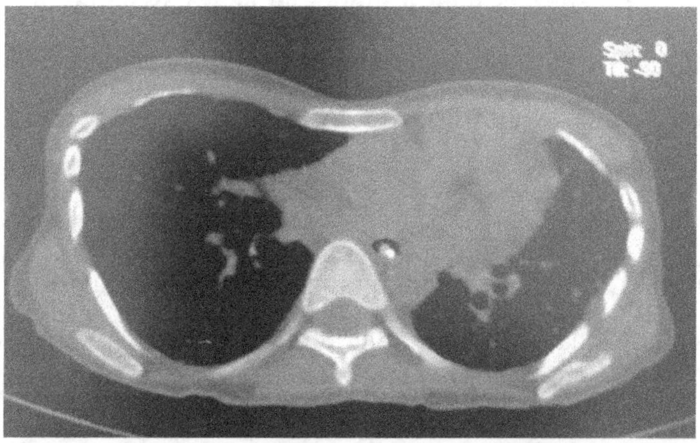

Figure no: 3.22

CT thorax axial image of a 45 years female patient suffering from pectus excavatum at Vertebral level T6 showing

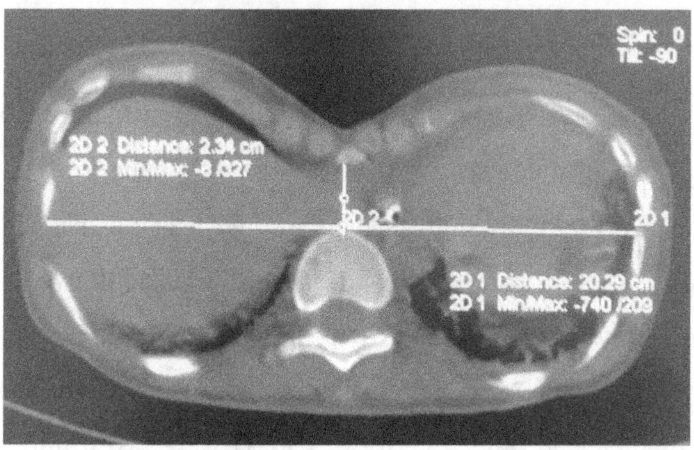

Figure no: 3.23

CT thorax axial image of a 45 years female patient suffering from pectus excavatum at Vertebral level T9 showing significantly reduced

Rather than radiographs, high-quality CT scans were used in this study to measure transverse and anteroposterior diameters in the axial plane. Despite the standard measurement of the chest wall with a digital caliper, the measurement tool used as the measurement methodology was the PACS built with CT machine. A previous study by Sverzellati N et al.[202], Cassart M et al.[203], and Arakawa H et al.[204] found it to be more accurate and better. Furthermore, previous studies only measured a specific unspecified vertebral level that varied between patients Haller J[78]. For a more accurate result, we measured chest dimensions at 10 vertebral levels from T3 to T12 in each normal control patient and patients suffering from chronic obstructive pulmonary disease (COPD) in case group. Although the Haller index was determined by Daunt et. al. in 2004[79] for describing chest dimensions in those who had no pectus excavatum chest anomaly. Haller indices were measured at 10 vertebral levels in normal pediatric subjects aged 8 to 18 years (James E. Archer)[77], they wanted to know the changes in the Haller indices at different vertebral levels and to predict the growth and development of the thorax. The overall mean Haller index was found 2.55 with a minimum and maximum variation of 2.26 to 3.30 for control males. The overall mean Haller index was found 2.60 with a minimum and maximum variation was 2.26 to 3.43 for normal female (controls). The overall mean Haller index was 2.00 with a minimum and maximum range between 1.49 to 2.26 for COPD male group. Chronic obstructive pulmonary disease (COPD) women had overall mean Haller index 1.99 with a minimum and maximum range of variation between 1.43 to 2.71. This shows that men had larger and more powerful chest cavities than women. Among COPD male and female, the deformation of thoracic cage was not significantly associated with age of the patient. Deformation of thoracic was more marked in patients with emphysematous lung than chronic bronchitis.

The overall results of the current study, which included both men and women, revealed increased anteroposterior diameter but no significant change in transverse diameter in COPD compared with normal controls. This finding was consistent with the findings of SJ Lim et al[117]. If the changes in rib cage dimensions were due to hyperinflation that was more related to severe airflow obstruction. The ribcage may then be subject to a greater structural change in patients with COPD. However, when the increase in anteroposterior diameter and the severity of COPD were taken into account. In the current study, there was no significant difference based on COPD severity. Patient age was found to be an important factor in chest changes in a normal control population.

The intervertebral disc space is typically reduced in older people of both sexes, resulting in a more convex backward curvature of the spine known as kyphosis Bartynski WS et. al.[205]. This shortening of the curvature of the spine results in narrow intercostal spaces and, as a result, a smaller rib cage (Sharma G et. al.)[206]. The current study found that anteroposterior was slightly increased in old age group (50-59 and 60-70 years, refer table no:16) as compared to early young adults and middle age group participants. However, the difference was statistically insignificant. Transverse diameters increased with increasing age of the patient and shortening of the spinal curvature. Anteroposterior diameter was more increased than transverse diameter. Thereby reduced value of Haller indices were reported in normal old (50-70 years) male and female patients. It is a known fact that COPD is more common in older people. Probably this can be the reason behind it. Consequently, age-related changes in the thorax were considered to be more important factors in these patients.

It is well known that the shape and dimensions of the ribcage differ according to gender. Many previous studies have investigated factors associated with changes in chest shape and found that gender has a greater effect on chest diameters than other factors such as BMI or age as reported by Sverzellati N et al.[202]. It is also known that the rib cage of women is 10 to 20% smaller than that of men, which is primarily due to the shorter transverse and anteroposterior (AP) diameter of the rib cage as compared to the men as reported by Bellemare F et. al.[207], Thurlbeck WM et. al.[176] and Weaver AA et. al.[208]. The results of present study were consistent with Bellemare F et. al.[207], Thurlbeck WM et. al.[176] and Weaver AA et. al.[208]. In our study, the proportion of normal male and female (controls) and COPD male and female (cases) were equal (96 in each group). Therefore, present study strongly favors the above finding.

However, in COPD male group, we found an overall increased average anteroposterior diameter (12.18 cm) as compared to normal male's average AP diameter (9.63cm) at p-value **<0.001**. In COPD female overall increased average anteroposterior diameter (10.90) as compared to normal female's average AP diameter (8.57cm) at p-value **<0.001.** Obliviously among COPD male and female increment was statistically significant as compared to normal group. Although among COPD group between male and female participants the difference of increased AP diameter was statistically insignificant. In COPD group average mean Haller index for male was 2.00 and for female was 1.99 (Ref table no:38). The average mean Haller index were nearly equal. This suggests that the sex factor was not associated with the increase in anteroposterior diameter in the COPD group among male and female. SJ Lim et al.[117] found a strong positive relationship between patient height and transverse diameter. Patient weight is related to anteroposterior diameter [117]. The current study had a limitation in that we did not measure the height and weight of patients in any group. The current study focused primarily on changes in the anteroposterior diameter in COPD, which revealed an increased extent of COPD. Salito et al.[209] used CT measurements to study five patients with severe emphysema. In their study in COPD group, it has been found that COPD patients had a larger anteroposterior diameter (Lando Y et.al.)[210] as compared to normal controls. Patients with severe COPD who have undergone lung volume reduction surgery may have a smaller anteroposterior chest diameter[210].

Because this study included lateral measurements, AP diameter, and Haller index for each vertebral level (10 levels) from T3 to T12 to determine the impact of COPD on the upper and lower chest. Also, how did the Haller index change and behave in COPD patients at different levels of the vertebral column compared to normal subjects? Findings of the present study showed that the transverse diameter was slightly smaller in COPD men compared with normal control men from T3 to T7. The lateral (transverse) diameter in COPD group for male was slightly higher in the lower part from T8 to T12. This shows that the shape of the chest has changed in the upper and lower regions of the chest in male. The same pattern was reported for T3 to T8 levels in normal control women as compared with women with COPD. The shape of the upper and lower thorax was different between males and females. COPD patients had a lower Haller index at each particular vertebral level than the normal controls. We also found that in the normal control group, the Haller index was higher in women than in men at each vertebral level except T3. In the COPD group, the pattern was changed slightly as the Haller index was higher in men than in women only at T4, T5, and T6. At rest, other levels were normal as in the control group. The Haller index was highest in both groups at the proximal level (Vertebral level-T3), then decreased until T11 and increased at T12 as compared to T11 reported by James E. Archer et. al.[77]. In normal group our finding was similar as reported by James E. Archer.

In lung transplantation, it is important to match the size of the donor and recipient organs. Eberlein M, et. al.[211] advocated changing the chest assignment mechanism from a height- or weight-based strategy to a size-matching process that uses complex estimates of limb size. Consequently, present study becomes more useful when combined with the weight, height, and spirometric parameters for lung size matching process for donor and recipient. Some other studies have been conducted to support height- or weight-based strategies for size-matching processes that use accurate estimates of organ size for lung transplantation as studied by Eberlein M and Diehl E et. al[212] and Eberlein M, Reed RM et. al.[213]. In size matching for organ for lung transplantation only height and weight of the patient are not whole sole parameter. Therefore, if transverse diameter, anteroposterior diameter, Haller index per vertebral level, height, weight, and spirometric parameters are considered together will give exact estimates of lung size of donor and recipient.

Before the 1970s, histological and post-mortem investigations were necessary to investigate lung structural alterations. The development of CT allowed noninvasive visualization of the thorax and assessment of lung anatomy. CT can detect major morphological aspects of COPD, such as emphysema, bronchial wall thickening, and air trapping. These pathologies contribute directly to airflow obstruction and, therefore, CT has the potential to provide vital insights into the underlying pathophysiological changes of COPD. Notwithstanding this, regular CT imaging is not commonly used in clinical practice, and its role in the treatment and research of COPD is not well established. The Global Initiative for COPD (GOLD) approach does not regularly advocate CT scanning in COPD and only suggests it for differential diagnosis or when surgical alternatives are being investigated (Global Initiative for Chronic Obstructive Lung Disease). A survey of opinions among respiratory doctors and surgeons in the United Kingdom found that just 32% believed it was important for patients with severe COPD to obtain a CT scan as reported by McNulty W et al 2014[214].

There were several reasons why CT has not become more widely used in COPD. One of the most significant was the perception that employing CT does not change management, which was not technically correct. Patients with advanced upper lobe emphysema and restricted exercise ability were found improved following lung volume reduction surgery, according to the NETT (National Emphysema Treatment Trial) studied by Fishman A et al 2003[215]. There has also been tremendous improvement and expertise in employing endobronchial therapies for severe emphysema including coils, valves, sealant, and airway bypass. These therapies can result in considerable improvements in lung physiology and functional ability in the right individuals with severe heterogeneous emphysema reported by Deslee et al 2014[216] and Sciurba FC et al 2010[217].

Computerized Tomography imaging can also detect bullous illness (emphysema) that was amenable to surgery, and bullectomy can result in better health (Palla et. al. 2005)[218]. As a result, it becomes evident that patients with severe COPD require Computerized Tomography imaging to assess who is eligible for these procedures. The current study was about to examine axial Computed Tomography thorax pictures of normal persons and COPD patients from North Indian population samples for structural alterations in the thoracic cage to discover answers to these issues. Measure radiographic maximum internal transverse (width) and minimum anteroposterior (height) diameters to determine the Haller index at each vertebral level, starting at the third thoracic (T3) vertebral to twelfth thoracic (T12) levels consecutively in the craniocaudal direction, applying Haller`s methodology in both normal subjects and patients of COPD.

Presently, the Global Strategy for the Diagnosis, Management, and Prevention of COPD (GOLD guidelines) does not utilize CT data during the first clinical diagnosis and relies solely on spirometry reported by Pauwels & Buist 2001[219]. However, enough scientific literature suggests that CT is an important and indispensable tool for the evaluation of COPD.

Chronic airflow obstruction can be caused by several disorders such as bronchiectasis, upper airway lesions, bronchiolar diseases, interstitial lung diseases, and so on, all of which can elicit clinical symptoms similar to COPD. High-resolution computerized Tomography (HRCT) can distinguish between several sources of airflow restriction. Kurashima et al. (2005)[220] demonstrated in 516 consecutive individuals with post bronchodilation FEV1/FVC was less than 70%. Computerized Tomography was capable of identifying 12.7% of patients with non-COPD lung illnesses such as sarcoidosis, diffuse panbronchiolitis, and pneumoconiosis. The precise identification of the underlying illness causing air flow blockage is critical in these patients' care and prediction of response to therapy.

Even in asymptomatic smokers with normal lung functioning, HRCT can identify pulmonary emphysema. Recent GOLD standards eliminated stage 0, which included asymptomatic patients who were smokers and had normal pulmonary functions; nonetheless, Sverzellati et al (2007)[202] discovered emphysema in 13/18 people with stage 0. It demonstrates that HRCT was a sensitive method for detecting emphysema before it manifests clinically or with abnormal lung functioning. Early detection of emphysema may be extremely valuable in preventing its development through smoking cessation and therapeutic procedures (Morgan 1992)[221].

Emphysema is typically characterized as an abnormal permanent expansion of air spaces (alveoli) distal to the terminal bronchioles, followed by alveolar wall disintegration. On CT, emphysematous alterations can be seen. Rosenblum et. al.[222] observed regions of decreased lung attenuation and mean lung density in individuals with COPD in 1978.

Emphysema, which has a parenchymal-predominant pathology in COPD, and chronic bronchitis, which has an airway-predominant pathology, are two separate phenotypes that may have developed due to different reactions to the pathogenesis caused by smoking (Patel et al 2006)[223]. Customarily, the emphysematous phenotype is linked with a more severe form of illness (Boschetto et al 2003, 2006)[224]. On High-Resolution Computed Tomography, COPD patients with emphysema had a higher BODE index (body mass index, airflow obstruction, dyspnea, exercise performance) and a lower inspiratory capacity to total lung capacity ratio (IC/TLC) than those without emphysema. This shows that people with emphysema have more widespread systemic inclusion (Boschetto et al 2006)[225]. The capacity to distinguish between airway-predominant and parenchymal-predominant disease in COPD may be beneficial in tailoring therapy to minimize airway remodeling and parenchymal deterioration (Grenier 2005)[226].

A CT scan can also reveal information about the lung lobes and zone-wise distribution of emphysema. The function of emphysema distribution on CT as a predictor of mortality is a current issue. One research found that a higher proportion of emphysema in the lower lungs was predictive of death; the authors theorized that lower-lobe emphysema might be either a sign of increasing disease severity or a phenotypic or pathobiological variation (Martinez et al 2006)[227]. Other research investigated the impact of emphysema distribution patterns on various lung function indicators. A larger proportion of emphysema in the core was linked to a lower diffusion capability (Aziz et al 2005)[228]. The contribution of emphysema to pulmonary dysfunction (as measured by FEV1/FVC%) in the core may be greater than at the periphery (Nakano et. al. 1999)[229].

During the last couple of decades, CT has been widely used to evaluate the emphysematous and chronic bronchitis components of COPD. The initial work by the Edinburg group was to use computed tomography to diagnose emphysema in living patients (Gould et. al. 1988)[230]. CT has now become routinely used to diagnose and quantify COPD and its subtypes. Several researchers have investigated the relationship between semi-quantitative COPD grading and CT, pathology, clinical characteristics, and pulmonary function testing (Sanders et al) 1988[231]; (Gupta et al) 2008[232], 2009[233]. Emphysema CT indices depict lung structure and were the most accurate approach to quantifying emphysema severity in real life (Newell et al 2004)[234]. The emergence of multidetector-row CT (MDCT) scanners has offered excellent capabilities to examine changes in both small and big airways. Furthermore, dynamic imaging by MDCT allows for the evaluation of perfusion and ventilation in the lung parenchyma. As a result, a CT scan precisely describes the disease process by giving accurate anatomic and functional data from the area of concern.

HALLER INDEX

According to Henry Gray, 2016[3] the transverse diameter is smaller at birth than it is in adulthood. The fraction of transverse diameter increases when a person learns to walk. Females have a lower absolute and proportionate thoracic capacity than males. Because the female sternum is shorter, the thoracic inlet is more sloped. The suprasternal notch is level with the third thoracic vertebra posteriorly in females. The thoracic inlet is less sloping in males because the sternum is longer, and the suprasternal notch is level with the second thoracic vertebra posteriorly studied by Harrison G A 1964[235]. The size of the thoracic cage varies constantly in all humans, based on rib and diaphragm movements during breathing and the degree of abdominal visceral distension (James E et al 2016)[77]. As a child grows older, the anteroposterior, transverse, and vertical dimensions of the thoracic cage increase correspondingly. It is still unknown how the thoracic cage develops in three dimensions. It is uncertain what are the common connection between transverse diameter and anteroposterior diameter at the most proximal and most distal thoracic levels reported by James E et al 2016[77]. The connection between transverse and anteroposterior diameters changes with age in North Indians, and male and female genders differ. Furthermore, it is uncertain how the connection between these two dimensions in North Indians evolves with developmental spurts and behaves at maturity.

Haller et. al.[78] devised a strategy for quantifying the connection between transverse and anteroposterior dimensions in pectus excavatum patients to categorize them as surgically treatable or non-interventional correctable. The Haller index was calculated by diving transverse and minimum anteroposterior internal thoracic diameters from an axial CT thorax image. This indicator was used to help patients suffering from pectus excavatum disease diagnosis and to make decisions for surgical option of the treatment. The measurement was collected from an axial CT thorax slice at the most severe level of deformation. The anteroposterior diameter was measured from the posterior surface of the sternum bone anteriorly to the anterior surface of the body of the thoracic vertebra posteriorly at the same location. Even though the sternum bone protrudes laterally concerning the vertebral column, the internal transverse and anteroposterior dimensions are measured in such a way that they intersect at a straight angle. Landmarks were traditionally thought to be skeletal rather than soft tissue above the bones. Daunt et al (Daunt S, et al 2004)[79] developed the Haller index to define thoracic measures for those who do not have a pectus excavatum thoracic cage anomaly. This time, each participant had one vertebral level assessed, and each person's vertebral level was different. James E. Archer et al. (James E et al 2016)[77] reported thoracic cavity dimensions in the British population using Haller's methodology. Is it relevant to other populations, such as north Indians, because it has been examined and researched less among Indians? Because Height, weight, body mass index (BMI), gender, genetic constitution, and physical body built and ethnicity of the north Indians have different from the other races of the world. As a result, it is necessary to conduct the study to determine the anticipated pattern of growth ratio per-vertebral thoracic level in diverse geographically distributed human populations concerning age and sex.

The Haller index was created and developed by Haller et al.[78]. This index was created by dividing the greatest internal transverse thoracic diameter by the lowest anteroposterior diameter as assessed on an axial CT thorax picture. This metric was used to evaluate pectus excavatum. This index was calculated and measured at a single vertebral level per subject. It was also obvious that females have greater Haller indices per vertebral level than males. One intriguing aspect is that the difference between mean Haller indices in adult to middle-aged males and females was not statistically significant per vertebral level in the present study. Previously, differences were important because, as people grow older, their growth rates and vertebral column curvature has changed. In adults and middle-aged people, the Haller index relates to the curvature of the spine, which can vary depending on the thickness (height) of the intervertebral disc. Our results through comparison of minimum anteroposterior diameter between normal and COPD groups among males and females showed a highly significant association ($p<0.001$) from T3-T12. The anteroposterior diameter from cranial vertebral level T3 to T11 was seen successively in increasing pattern and again decreased value at T12 as compared with T11. Among males and females highest value Haller index was reported at T3 among both normal and COPD groups. Haller index also showed decreasing pattern from T3 to T11 and increased value at T12 as compared to T11 was found. When comparison was made for Haller index between males and females among normal group, it was seen that females always had higher value of Haller indices per vertebral level. Although, the difference was not statistically significant because the growth rate remains constant in early adult hood to middle aged (18-49) people.

When Comparison of minimum anteroposterior diameter between males and females among COPD group was made, smallest anteroposterior diameter was found again at cranial vertebral level T3. Anteroposterior diameter showed an increasing pattern from T3 to T11 and again decreased value at T12 as compared to T11 in COPD group among males and females. Vertebral level T11 had highest value. However, anteroposterior diameter was significantly higher in COPD males and females as compared to normal males and females. Maximum Haller index was found at T3 with decreasing trend till T 8 and slight increase at T9 in COPD males. Vertebral level T9, T10 and T11 nearly had same values of Haller index in COPD males and T12 had comparatively higher value as compared to T11. The COPD males had significantly lower values of Haller index at each vertebral level as compared to normal males. Same was observed for COPD females. When a comparison of the Haller index was made in various age groups among normal group maximum Haller index was found at T3 with no significant association. Although, comparatively lower values of Haller index were found in old age groups 50-59 and 60-70 years. Similarly in the COPD group maximum Haller index was seen at T3 and a significant association was seen at T7 ($p=0.016$). Among COPD group patients belonging to any age group had always lower value of Haller index per vertebral level than normal group. But the pattern was same that cranial level had always higher value than caudal. When age groups 18 – 29 years, 30 – 39 years, 40 –49 years, and 50-59 years were analyzed separately between normal and COPD groups a highly significant association ($p<0.001$) was seen at all stages with the maximum at T3.

Biavati M et al 2020[236] discovered that radiographic measurements of pectus patients were more prevalent in females than males and that women had a higher frequency of pectus than men. In present study we also reported a single patient of pectus excavatum who was 45 years female. Its prevalence in present study was found 0.26%. The Haller index and Correction indices are related to height and weight in the general population, regardless of age, gender, or ethnicity. Over a period of seven years, repeat imaging of a subset of patients (n = 992) revealed declines in the mean Haller index and Correction indices, indicating a shift to a more circular axial thorax, with less sternal depression, over time. Pectus excavatum is the most common anterior chest wall deformity, which had been estimated to have an incidence of one (1) in forty (40) to one in four hundred (400) people across different cohorts (Westphal FL et. al, Coskun ZK et.

al.)[237,238]. In clinical practice it was usually observed that males were referred for evaluation 3-5 times more often than females reported by Cartoski MJ et. al.[239].

Prasad H et. al.[240] found results that T3 has the highest mean Haller index in males. It declined from T3 to T8, then slightly increased at T9, then dropped again at T10 and T11, and finally increases again at T12. T3 was shown to have the highest Haller index among females. It declines from T3 to T8. Vertebral level T9 showed a little rise, T10, and T11 show a drop, and T12 showed an increase. Females had higher Haller indices per vertebral level than males.

Archer JE et al 2016[77] discovered that the Haller index was highest in both males and females at the first measurable level of T3, then drops to the lowest value at T8 and T9 before increasing somewhat in the more caudal vertebral levels. The Haller index rises with age in both sexes, especially in the more cranial thorax and, to a lesser extent, the more caudal thorax. This is because of a higher rise in thoracic breadth compared to thoracic height. With advancing age, the mid-thorax shows a minimal change. There is some variance between the sexes, with the increase in the Haller index largest in females. In both sexes there is a slowdown in the rise of the index, noted at age 14 in men and 15 in females. It is plausible to believe that this slowing is due to the teenage growth spurt.

S. Allwyn Joshua et al 2014[241] indicated that Haller's index in pre-pubertal exhibited a negligible range with an average of 2.2, but the post-pubertal group displayed an extensive range of 1.6 to 3.1 with an average of 2.37. Sterno-vertebral distance in the post-pubertal group showed a mean of 100.4±17 ($p<0.05$) with a range of 59 to 140 in comparison to the pre-pubertal value of 79.7±14 (59 to 98) and the transverse distance in post-pubertal also showed significantly higher value of 230.2±26 ($p<0.05$), while pre-pubertal group showed an average of 172±32. It was thus concluded that alterations in the geometric dimensions and shape of the thoracic wall, change in the pattern and the mechanism of respiration were influenced not only due to change in the inclination of the rib and the curvature of the vertebral column but also due to the position of the sternum plays a pivotal role.

Our results-based age distribution of cases and Controls illustrated that in the normal group, the maximum number of subjects were found in the age group 18-29 years while in the COPD group, the maximum number of subjects were 68 (constituted 35.04%) found in 60 - 70 years. The mean age of normal group participants was 42.34 with SD±14.77 years. Mean age of COPD participants was 51.08 with SD±13.46 years. Lim SJ et. al.[117] found that the COPD group which was comprised of 76 males, and 9 females with mean age, of 70.6±7.1 years. While his normal group was comprised of 23 males and 7 females with mean age, of 69.9±4.2 years. Viswambhar V et. al.[108] found that among COPD group the mean age was 60.04 ± 3.58 with (51.67%) male and 18 (45%) being female and among the healthy controls, the mean age was 59.41 ± 3.18 with 48.33% male and remaining 55% were female.

Results also showed a significant association in both the male COPD and control group with T4 (p=0.044), T5 (p=0.028), and T6 (p=0.037). The results regarding minimum anteroposterior diameter among male control and cases were also significant with T3-T12 (p<0.001). Among female control (normal) and cases (COPD), comparison of maximum internal transverse diameter showed a significant association with T3 (p=0.008), T4 (p=0.002), T5 (p=0.001), T7 (p=0.001), and T8 (p=0.021) and a highly significant association with T6 (p<0.001). Concerning minimum anteroposterior diameter among females in control & cases a significant association was found in both the group with T3-T12 (p<0.001). Results seen when males were compared to females concerning maximum internal transverse diameter and minimum anteroposterior diameter showed a highly significant association from T3 –T12 (p<0.001). Prasad H et al found results similar to our result such that the p-value for T3-T12 in mean transverse and AP diameters, as well as the maximum and minimum variability per vertebral level, in 73 male patients and 35 female patients, was p<0.001.

Careful donor quality assessment and size match can impact long-term survival in lung transplantation. If the donor lung quality and size match is satisfactory, the offer is tentatively accepted by that institute following confirmation of an appropriate virtual cross-match. The donor risk profile is thoroughly reviewed with the recipient, and after the recipient agrees, the institution dispatches a procurement team to the donor hospital, where the final acceptance is determined following an intraoperative examination.

To evaluate the donor's quality, the following factors are considered: the donor's history of current admission, past medical history, past surgical history, cause of death, the most recent chest x-ray and CT scan, bronchoscopy, sputum cultures, and most recent blood reports, as well as arterial blood gases, as described in the "Donor management protocol" article of this special issue (Edward C)[242].

Pioneering surgeons were more worried about surgical anastomoses and infection in the early years of lung transplantation than with the probable significant discrepancy of lung volume between the recipient and donor (Mallory GB et. al.)[243]. As a result, in the 1980s, the concept of matching donor and recipient lungs received little or no attention (Reitz BA et al)[244]. The American Thoracic Society lung transplant workshop in 1992 suggested utilizing allografts with "identical total lung capacity" between donor and recipient (Goldstein et al, Patterson GA et al)[245].

The chest cavity size in COPD patients is substantially greater than in normal people. Implanting a much bigger lung allograft may not result in the normalization of respiratory mechanics and may have a detrimental influence on physical endurance or 6-minutes' walk distance post-transplant and survival (Mason DP et. al.)[246.] Mason et al. showed that the utilization of extremely tiny (donor's predicted TLC (pTLC) less than 67% of recipient's actual TLC (aTLC)) or extremely big (donor's pTLC higher than recipient's aTLC) donor lungs in emphysema patients leads in worse survival with bilateral lung transplantation (Mason DP et al)[246].

Tamm et al. studied pre-operative TLC of 82 patients who underwent heart-lung transplantation (HLTx). They discovered that independent of the underlying condition, preoperative TLC, or donor pTLC, the postoperative TLC reached the recipient's pTLC following HLTx (Tamm et al). Furthermore, Chacon et al. studied patients who received either HLTx ($n = 6$) or bilateral LTx ($n = 9$) and found that postoperative TLC ($r = 0.724$, $p = 0.002$) was substantially connected with pTLC but not with preoperative TLC ($r = 0.148$, p = not significant) (Chacon RA et al)[247].

Larger excision of the lungs might result in the implantation of smaller lungs, which could explain why thoracic cavity volume (TCVs) is smaller in individuals with the restrictive lung disease. Some studies have shown that an enlarged graft might delay bronchiolitis obliterans syndrome and improve survival rates. Furthermore, Eberlein et al found that a larger graft was not related to an increased risk of postoperative problems. Similarly, they detected a favorable connection between postoperative 1-year TCV and pulmonary function test (PFT). These findings point to a link between the restricting group's preference for smaller organs and poorer results (Eberlein M et al)[211].

The National Emphysema Treatment Trial (NETT) showed that lung volume reduction (LVR) surgery might enhance lung function, exercise capacity, and survival in a subset of COPD patients (Geddes D et al and Fishman A et al)[248, 249]. The diaphragm flattens and straightens in COPD, resulting in asynchronous movement of the chest wall (Gilmartin JJ et al)[250]. Zoumot et al.[251] analyzed 26 patients who received lung volume recruitment (LVR) [surgical (n = 9) or bronchoscopic (n = 7)] or a sham/unsuccessful bronchoscopic therapy (control subjects, n = 10) (Zoumot Z et al)[251].

TCV was favorably linked with height. Taller people are more likely to have a higher TCV before developing lung illness and may have a higher TCV following LTx. BMI, on the other hand, was adversely associated with TCV. Obese people are known to have the lower functional residual capacity (FRC) (Jenkins SC et al and Boriek AM et al)[252, 253]. Obesity alters the form of the diaphragm, according to one study, and the diaphragm was moved cranially in obese patients (Boriek AM et al)[253]. Those findings were congruent with the findings of a different investigation, which found that BMI was adversely linked with diaphragm height as evaluated by posteroanterior chest radiography (Bellemare JF, et al)[254].

SUMMARY AND CONCLUSIONS

More study is needed to prove quantitative CT analysis as a disease imaging biomarker that may be utilized effectively to manage COPD patients and as a helpful research tool to provide insights into the illness. Technically, CT methods must be standardized since radiation dose, slice thickness, and reconstruction techniques may all affect the analysis. Methods for assessing numerous morphological aspects, particularly airway measurements and gas trapping, require improvement. More advanced tools may be required that not only quantify outcomes but also account for additional factors such as emphysema subtypes and airway wall attenuation. Extrapulmonary manifestation detection measures are also in the early stages of development, and further study is needed to improve them.

CT analysis must be compared to outcome measurements and disease activity indicators. There have been few long-term investigations so far, and the findings have been mixed. Studies must be undertaken to follow CT changes over time to provide an accurate picture of illness progression and how efficient CT imaging is in capturing this. There have also been few studies that look at quantitative CT indicators of illness and markers of inherent biological activity, such as inflammatory markers, microbiology, proteases, and other biomarkers. Our work demonstrates how this might be accomplished by combining bronchoscopic samples and CT imaging to establish a potentially significant link between proteases and small-airway disease in COPD. Similar techniques can be utilized to expand our knowledge of the active disease processes and endotypes happening in different compartments and areas of the lung, hopefully permitting the development of innovative targeted therapeutics.

Improving CT technology will also enhance its use in COPD. A primary limiting element is the radiation dosage and accompanying cancer concerns. Dose reduction measures are crucial if longitudinal scans are to be considered. Dual-energy CT technology is being developed and will have further uses, including functional imaging of the pulmonary vasculature. CT imaging might potentially be complemented with other imaging technologies to offer more information about disease activity. CT and MRI might be co-registered to integrate structural and functional imaging.

One of the most pressing issues is how to use CT analysis to designate clinically significant phenotypes. It is unclear that just subgrouping individuals into emphysema and airway-predominant illness would be sufficiently useful to guide treatment options. Cluster analysis and other statistical modeling approaches may yield intriguing results, but they are more likely to be research tools with little clinical applicability. Emphysema and bronchial wall thickening on CT have already been related to higher mortality and improving this information and integrating it with clinical data may help strengthen multidimensional prognostic tools such as the BODE index.

This method might have a substantial influence on COPD clinical trials and medication development. Quantitative CT has already been utilized as an outcome measure for the advancement of emphysema in 1-ATD, and this might be extended to medications that target airway inflammation and small-airway disease. The disease's variability presents considerable hurdles in clinical trial design. Because most trials include unselected patient groups, they must recruit a large number of people and endure a long period to discover enough responders. Improving understanding of patient heterogeneity is critical for improving pharmacological trials, and quantitative CT may play a role. CT might potentially be utilized to stratify trial participants into phenotypic groups that are more likely to react to the experimental medicine. CT might be utilized to identify which individuals have airway thickening/inflammation, and this subject group could be employed in any interventional investigation. Improving the stratification of research individuals may boost treatment effects and minimize variability in response, hence improving the likelihood of innovative medications being produced. This, in turn, will lead to more therapeutic uses for CT imaging in assessing which patients could benefit from innovative pharmacological drugs.

Bibliography

1. Henry Gray edited by Peter L. Williams. The Anatomical basis of medicine and surgery. Churchil Living Stone; 1995, 38 editions, p. 537.
2. T. W. Sadler edited by Jan Langman. Langman`s Medical Embryology. Walters Kluwer Pvt. Ltd, New Delhi; 2016, Chapter 6, page 71.
3. Henry Gray edited by Susan Standring. The Anatomical Basis of Clinical Practice. Elsevier limited; 2016, 41 editions, Chap 51, p. 905.
4. T. W. Sadler edited by Jan Langman. Langman`s Medical Embryology. Walters Kluwer Pvt. Ltd, New Delhi; 2016, Chapter 14, page 218.
5. Henry Gray edited by Susan Standring. The Anatomical Basis of Clinical Practice. Elsevier limited; 2016, 41 editions, Chap 51, p. 906.
6. Andreas V. The bones of the thorax. In: Garrison DH, Hast MH, editors. The Fabric of the Human Body. An Annotated Translation of the 1543 and 1555 Editions. Ch. 19. Northwestern University; 2003.
7. T. W. Sadler edited by Jan Langman. Langman`s Medical Embryology. Walters Kluwer Pvt. Ltd, New Delhi; 2016, Chapter10, page154.
8. Inderbir Singh edited by V Subhadra Devi. Human Embryology. Jaypee Brothers Medical Publishers; 2021, 12 editions, Chapter13, page184-85.
9. Erdil H, Yildiz N, Cimen M. Congenital fusion of cervical vertebrae and its clinical significance. Journal of Anatomical Society of India, 2003; 52(2): 125-127.
10. Inderbir Singh edited by V Subhadra Devi. Human Embryology. Jaypee Brothers Medical Publishers; 2021, 12 editions, Chapter13, page180-81.
11. T. W. Sadler edited by Jan Langman. Langman`s Medical Embryology. Walters Kluwer Pvt. Ltd, New Delhi; 2016, Chapter10, page 153.
12. Inderbir Singh edited by V Subhadra Devi. Human Embryology. Jaypee Brothers Medical Publishers; 2021, 12 editions, Chapter13, page184.
13. Goldstein L, Waugh T. Classification and terminology of scoliosis. Clin Orthop Rel Res 1973; 93:10–22.
14. Guille J, Forlin E, Bowen R. Congenital kyphosis. Orthop Rev 1993;235–9.

15. Bullough P, Boachie-Adjei O. Atlas of Spinal Disease. Philadelphia: Lippincott; 1988.
16. Scheuermann H. Röntgenologic studies of the origin and development of juvenile kyphosis, together with some investigations concerning the vertebral epiphyses in man and animals. Acta Orthop Scan 1934; 5:161.
17. Schmorl G. Über die an den Wirbelbandscheiben vorkommenden Ausdehnungs- und Zerreisungsvorgänge und die dadurch an ihnen und der Wirbelspongiosa hervorgerufenen Veränderungen. Verh Dtsch Pat Ges 1927; 22:50.
18. Zorab P. Respiratory tract disease chest deformities. BMJ1966;7 May:1155–6.
19. J.R. Cobb. Outline for the study of scoliosis. Am Acad Orthop Surg Instr Course Lect, 5 (1948), pp. 261-275.
20. Goldstein L, Waugh T. Classification and terminology of scoliosis. Clin Orthop Rel Res 1973; 93:10–22.
21. Drummond D, Rogola E, Gurr J. Spinal deformity: natural history and the role of school screening. Orth Clin North Am1979;10(4):751–9.
22. James J. The etiology of scoliosis. J Bone Joint Surg1970;52(3):410–9.
23. Seaver LH, Boyd E. Spondylocarpotarsal synostosis syndrome and cervical instability PubMed Am J Med Genet, 2000; 91(5): 340-344.
24. Henary Gray edited by Peter L. Williams. The Anatomical basis of medicine and surgery. Churchil Living Stone; 1995, 38 editions, p. 537.
25. Jit I, Bakshi V. Incidence of sterna foramina in North India. J Anat soc Ind, 1984, 33: 77-84.
26. Standering S, The anatomical basis of clinical practice. Gray's Anatomy, 41 editions, 53chapter, 2016;934.
27. Standering S, The anatomical basis of clinical practice. Gray's Anatomy, 41 editions, 53chapter, 2016;932.
28. Gabriel A, Donnelly J, Kuc A, Good A, Doros G, MatsuzP,et al.Ectopiacordis: A rare Congenital Anomaly. Clin Anat.2014; 27:1193-1199.
29. Ashley GT, The relationship between the pattern of ossification and the definitive shape of the mesosternum in man. J Anat. 1956; 90:87-105.

30. Neviere R, Montaigne D, Benhamed L, et al. Cardiopulmonary response following surgical repair of pectus excavatum in adult patients. Eur J Cardiothorac Surg. 2011;40(2): e77-e82. doi: 10.1016/j.ejcts.2011.03.045
31. Chung CS, Myrianthopoulos NC. Factors affecting risks of congenital malformations, I: analysis of epidemiologic factors in congenital malformations. Report from the Collaborative Perinatal Project. Birth Defects Orig Artic Ser. 1975;11(10):1-22.
32. . Shamberger RC. Congenital chest wall deformities. Curr Probl Surg. 1996;33(6):469-542.
33. Ma IT, Rebecca AM, Notrica DM, McMahon LE, Jaroszewski DE. Pectus excavatum in adult women: repair and the impact of prior or concurrent breast augmentation. Plast Reconstr Surg. 2015; 135(2): 303e-312e. doi :10 . 1097/ PRS.0000000000000990.
34. Colombani PM. Preoperative assessment of chest wall deformities. Semin Thorac Cardiovasc Surg. 2009;21(1):58-63. doi: 10. 1053/ j. semtcvs. 2009. 04.003.
35. Neviere R, Montaigne D, Benhamed L, et al. Cardiopulmonary response following surgical repair of pectus excavatum in adult patients. Eur J Cardiothorac Surg. 2011;40(2): e77-e82. doi: 10.1016/j.ejcts.2011.03.045.
36. Chao CJ, Jaroszewski DE, Kumar PN, et al. Surgical repair of pectus excavatum relieves right heart chamber compression and improves cardiac output in adult patients–an intraoperative transesophageal echocardiographic study. Am J Surg. 2015;210(6):1118-1125. doi: 10.1016/j.amjsurg.2015.07.006
37. Sigalet DL, Montgomery M, Harder J, Wong V, Kravarusic D, Alassiri A. Long term cardiopulmonary effects of closed repair of pectus excavatum. Pediatr Surg Int. 2007;23(5):493-497. doi:10.1007/s00383-006-1861-y.
38. Creswick HA, Stacey MW, Kelly RE Jr, et al. Family study of the inheritance of pectus excavatum. J Pediatr Surg. 2006;41(10):1699-1703. doi: 10. 1016/ j. jpedsurg. 2006.05.071.

39. Kotzot D, Schwabegger AH. Etiology of chest wall deformities—a genetic review for the treating physician. J Pediatr Surg. 2009;44(10):2004-2011. doi: 10.1016/j.jpedsurg.2009.07.029.
40. Koumbourlis AC. Pectus deformities and their impact on pulmonary physiology. Paediatr Respir Rev. 2015;16(1):18-24. doi: 10. 1016/ j. prrv. 2014.10.009.
41. Mak SM, Baludin BN, Naaseri S, Di Chiara F, Jordan S, Pandey S. Imaging of congenital chest wall deformities. Br J Rdiol, 2016; 89: 20150595.
42. Kelly RE Jr, Shamberger RC, Mellins RB, et al. Prospective multicenter study of surgical correction of pectus excavatum: design, perioperative complications, pain, and baseline pulmonary function facilitated by internet-based data collection. J Am Coll Surg. 2007;205(2):205-216. doi: 10.1016/j.jamcollsurg.2007.03.027.
43. Creswick HA, Stacey MW, Kelly RE Jr, et al. Family study of the inheritance of pectus excavatum. J Pediatr Surg. 2006;41(10):1699-1703. doi: 10.1016/j.jpedsurg.2006.05.071.
44. . Kotzot D, Schwabegger AH. Etiology of chest wall deformities—a genetic review for the treating physician. J Pediatr Surg. 2009;44(10):2004-2011. doi: 10.1016/j.jpedsurg.2009.07.029.
45. Robicsek F, Daugherty HK, Mullen DC, Harbold NB Jr, Hall DG, Jackson RD, et al. Technical considerations in the surgical management of pectus excavatum and carinatum. Ann Thorac Surg. 1974;18(6):549-64.
46. Coelho MS, Stori WS, Pizarro LD, Zanin AS, Gonçalves JL, Bergonse Neto N. Pectus Excavatum/Pectus Carinatum: Tratamento Cirúrgico. Rev Col Bras Cir. 2003;30(4):249-61.
47. Coelho MS, Santos A, Pizarro L, Zampier J, Krieguer R. "Pectus excavatum /pectus carinatum": tratamento cirúrgico. J Pneumol. 1983;10(Supl):47.
48. Fonkalsrud EW, Anselmo DM. Less extensive techniques for repair of pectus carinatum: the undertreated chest deformity. J Am Coll Surg. 2004;198(6):898-905.

49. Coelho MS, De Campos JR, Stori WS. Deformidades Congênitas da Parede Torácica Anterior. In: Saad Jr R, Carvalho WR, Ximenes Netto M, Forte V, editores. Cirurgia Torácica Geral. São Paulo: Editora Atheneu; 2005.p.765-91.

50. Coelho MS, Guilherme EV, Kume MK Vialle LA. Incidência de deformidades torácicas entre escolares de Curitiba. J Pneumol. 1982;9(Supl):175.

51. Lodi R, Bondioli A, Curti L, Bruni GC, Palmieri B. [Surgical correction of the pectus excavatum and carinatum in the adult. Report of an unusual case of combination of the straight back and pectus excavatum] Minerva Chir. 1975;30(3):131-8.

52. Chidambaram B, Mehta AV. Currarino-Silverman syndrome (pectus carinatum type 2 deformity) and mitral valve disease. Chest. 1992;102(3):780-2.

53. Fokin AA, Steuerwald NM, Ahrens WA, Allen KE. Anatomical, histologic, and genetic characteristics of congenital chest wall deformities. Semin Thorac Cardiovasc Surg 2009; 21: 44–57. doi: http://dx.doi.org/ 10.1053/j.semtcvs.2009.03.001.

54. Marfan A B. Un cas de deformation congenitale des quatre membres, plus prononce aux extremetes, characterise par l'allongement des os avec uncertain degree d'amincissement. Bull Mem Soc Med Hop (Paris) 189613220–226.

55. Judge D P, Dietz H C. Marfan's syndrome. Lancet 20053661965–1976.

56. Alan G S, Andrew W. Marfan's syndrome and the heart. Arch Dis Child. 2007; 92:351–356. doi: 10.1136/adc.2006.097469.

57. Stuart AG, Wilson DG. Marfan syndrome: is early diagosis useful? Curr Paediatr 1998; 8:121–5.

58. DePaepe A, Devereux RB, Dietz H, et al. Revised diagnostic criteria for the Marfan syndrome. Am J Med Genet 1996; 62:417–26.

59. Chetty SP, Shaffer BL, Norton ME. Management of pregnancy in women with genetic disorders, Part 1: Disorders of the connective tissue, muscle, vascular, and skeletal systems. Obstet Gynecol Surv. 2011 Nov;66(11):699-709. [PubMed]

60. Malfait F, Wenstrup RJ, De Paepe A. Clinical and genetic aspects of Ehlers-Danlos syndrome, classic type. Genet Med. 2010 Oct;12(10):597-605. [PubMed]
61. Malfait F, Francomano C, Byers P, Belmont J, Berglund B, Black J et al. The 2017 international classification of the Ehlers-Danlos syndromes. Am J Med Genet C Semin Med Genet. 2017 Mar;175(1):8-26. [PubMed]
62. De Paepe A, Malfait F. The Ehlers-Danlos syndrome, a disorder with many faces. Clin Genet. 2012 Jul;82(1):1-11. [PubMed]
63. J G Ayres, F M Pope, J F Reidy, T J Clark. Abnormalities of the lungs and thoracic cage in the Ehlers-Danlos syndrome. BMJ. April 1985; 40:300-305.
64. Roberts AE, Allanson JE, Tartaglia M, Gelb BD. Noonan syndrome. Lancet 2013; 381: 333-42.
65. Binder G, Grathwol S, von Loeper K, et al. Health and quality of life in adults with Noonan syndrome. J Pediatr 2012; 161: 501-5. e1.
66. Mendez HM, Opitz JM, Reynolds JF. Noonan syndrome: a review. Am J Med Genet 1985; 21: 493-506.
67. Alireza T, Peyman E, Zahra KK, Mohammadreza A. Noonan syndrome – a new survey.AMS. 2017; Volume 13, Number 1, pages 215–222.
68. Tartaglia M, Kalidas K, Shaw A, et al. PTPN11 mutations in Noonan syndrome: molecular spectrum, genotype-phenotype correlation, and phenotypic heterogeneity. Am J Hum Genet 2002; 70: 1555-63.
69. Lepri F, De Luca A, Stella L, et al. SOS1 mutations in Noonan syndrome: molecular spectrum, structural insights on pathogenic effects, and genotype-phenotype correlations. Hum Mutat 2011; 32: 760-72.
70. Marino B, Digilio MC, Toscano A, Giannotti A, Dallapiccola B. Congenital heart diseases in children with Noonan syndrome: an expanded cardiac spectrum with high prevalence of atrioventricular canal. J Pediatr 1999; 135: 703-6.
71. Bhambhani V, Muenke M. Noonan syndrome. Am Fam Phys 2014; 89: 37-43.

72. Naficy S, Shepard NT, Telian SA. Multiple temporal bone anomalies associated with Noonan syndrome. Otolaryngol Head Neck Surg 1997; 116: 265-7.
73. Qiu WW, Yin SS, Stucker FJ. Audiologic manifestations of Noonan syndrome. Otolaryngol Head Neck Surg 1998; 118: 319-23.
74. Romano AA, Allanson JE, Dahlgren J, et al. Noonan syndrome: clinical features, diagnosis, and management guidelines. Pediatrics 2010; 126: 746-59.
75. Henary Gray edited by Peter L. Williams. The Anatomical basis of medicine and surgery. Churchill Living Stone; 1995, 38 edition, chap 4, p. 366.
76. Harrison G A, Weiner J S, Tanner J M, Barnicot NA. Human biology. Clarendon press: Oxford, 1964, chap 19.
77. James E. Archer, Adrian G., Fiona B. and Paul P. The measurement of the normal thorax using the Haller index methodology at multiple vertebral levels. J Anat., 2016; V 229, p 577- 581.
78. Haller J, Kramer S, Leitman S (1987) Use of CT scans in selection of patients for pectus excavatum surgery: a preliminary report J Pediatr Surg 22, 904–906.
79. Daunt S, Cohen J, Miller S(2004) Age-related normal ranges for the Haller index in children..Pediatr Radiol34, 326–330.
80. Kasper DL, Fauci AS, Hauser S, et al, editors. Harrison`s Principle of Internal Medicine. Mc Graw Hill Education; 2018, 20 editions, pages1943.
81. Kasper DL, Fauci AS, Hauser S, et al, editors. Harrison`s Principle of Internal Medicine. Mc Graw Hill Education; 2018, 20 editions, pages1990.
82. Kasper DL, Fauci AS, Hauser S, et al, editors. Harrison`s Principle of Internal Medicine. Mc Graw Hill Education; 2018, 20 editions, page1995.
83. Gilmartin JJ, Gibson GJ. Abnormalities of chest wall motion in patients with chronic airflow obstruction. *Thorax.* 1984;39:264–71. [PMC free article] [PubMed] [Google Scholar].
84. Hoover CF. The diagnostic significance of inspiratory movements of the costal margins. Am J Med Sci. 1920; 159:633–46. [Google Scholar]
85. Davidson, Christopher R.W. Edwards, Ian A.D., Christopher Hansleit, Edwin Chilvers. Davidsons`s Principles and Practice of Medicine. Churchill Livingstone. 1995; Editions 17, Chapter 6, pages332-333.

86. Davidson, Christopher R.W. Edwards, Ian A.D., Christopher Hansleit, Edwin Chilvers. Davidsons`s Principles and Practice of Medicine. Churchill Livingstone. 1995; Editions 17, Chapter 6, pages334-335.
87. Davidson, Christopher R.W. Edwards, Ian A.D., Christopher Hansleit, Edwin Chilvers. Davidsons`s Principles and Practice of Medicine. Churchill Livingstone. 1995; Editions 17, Chapter 6, pages338-339.
88. Barker AF. Bronchiectasis. N Engl J Med. 2002; 346:1383–1393.
89. King PT, Holdsworth SR, Freezer NJ, Villanueva E, Holmes PW. Characterisation of the onset and presenting clinical features of adult bronchiectasis. Respir Med. 2006; 100:2183–2189.
90. McGuinness G, Naidich DP. CT of airways disease and bronchiectasis. Radiol Clin North Am. 2002; 40:1–19.
91. Rennard SI, Drummond MB. Early chronic obstructive pulmonary disease: definition, assessment, and prevention. Lancet. 2015; 385:1778-88.
92. Rennard SI. COPD: overview of definitions, epidemiology, and factors influencing its development. Chest. 1998; 113:235S-41S.
93. Lopez AD, Shibuya K, Rao C, Mathers CD, Hansell AL, Held LS, et al. Chronic obstructive pulmonary disease: current burden and future projections. Eur Respir J. 2006;27:397-412.
94. Petty TL. COPD in perspective. Chest. 2002; 121:116S-20S.
95. Dyspnea. Mechanisms, assessment, and management: a consensus statement. American Thoracic Society. Am J Respir Crit Care Med. 1999; 159:321-40.
96. Kilburn KH, Amundsen T. Anteroposterior chest diameter in emphysema. From maxim to measurement. Arch Intern Med. 1969; 123:379-82.
97. Walsh JM, Webber CL, Jr., Fahey PJ, Sharp JT. Structural change of the thorax in chronic obstructive pulmonary disease. J Appl Physiol (1985). 1992; 72:1270-8.
98. Gilmartin JJ, Gibson GJ. Abnormalities of chest wall motion in patients with chronic airflow obstruction. Thorax. 1984; 39:264-71.

99. Gilmartin JJ, Gibson GJ. Mechanisms of paradoxical rib cage motion in patients with chronic obstructive pulmonary disease. Am Rev Respir Dis. 1986; 134:683-7.

100. Aliverti A, Quaranta M, Chakrabarti B, Albuquerque ALP, Calverley PM. Paradoxical movement of the lower ribcage at rest and during exercise in COPD patients. Eur Respir J 2009; 33:49-60.

101. Hogg JC, Chu F, Utokaparch S, Woods R, Elliott WM, Buzatu L, et al. The nature of small-airway obstruction in chronic obstructive pulmonary disease. N Engl J Med. 2004; 350:2645-53.

102. Kim V, Davey A, Comellas AP, Han MK, Washko G, Martinez CH, et al. Clinical and computed tomographic predictors of chronic bronchitis in COPD: a cross sectional analysis of the COPD Gene study. Respir Res.2014; 15:52.

103. McDonough JE, Yuan R, Suzuki M, Seyednejad N, Elliott WM, Sanchez PG, et al. Small-airway obstruction and emphysema in chronic obstructive pulmonary disease. N Engl J Med. 2011; 365:1567-75.

104. Surendran S, Mohan A, Valamparampil MJ, Nair S, Balakrishnan SK, Laila AA, Reghunath R, Jose C, Rajeevan A, Vasudevakaimal P, Surendrannair AT, Nujum ZT, Varghese S, Mohan A. Spatial analysis of chronic obstructive pulmonary disease and its risk factors in an urban area of Trivandrum, Kerala, India. Lung India. 2022 Mar-Apr;39(2):110-115.

105. Feizi, H., Alizadeh, M., Nejadghaderi, S.A. *et al.* The burden of chronic obstructive pulmonary disease and its attributable risk factors in the Middle East and North Africa region, 1990–2019. *Respir Res* 23, 319 (2022). https://doi.org/10.1186/s12931-022-02242-z.

106. Chen, Y., Li, J., Dong, B. *et al.* Two-dimensional shear wave elastography: a new tool for evaluating respiratory muscle stiffness in chronic obstructive pulmonary disease patients. *BMC Pulm Med* 22, 441 (2022). https://doi.org/10.1186/s12890-022-02231-4.

107. Chavan SD, Vijay S Patil, Basanagouda K Patil. Comparison of thoracic dimensions of chronic obstructive pulmonary disease patients and healthy

controls: A cross-sectional study. Int J Adv Res Med 2021; 3(1):75-77. DOI: 10.22271/27069567.2021.v3.i1b.105.

108. Viswambhar V, Bandlamudi S, Aruna S, Meenakshi NS, Ashwin K J K, Navaneetham K. An Analytical Cross-Sectional Study Comparing Thoracic Dimensions between Chronic Obstructive Pulmonary Disease (COPD) Patients and Healthy Controls. IJCMR. Feb 2020; Vol7, Issue2, page B5-B8.
109. Thiago Krieger Bento da Silva, Matheus Zanon, Stephan Altmayer, Gabriel Sartori Pacini, Guilherme Watte, Renato Stein et al. High-resolution CT pulmonary findings in children with severe asthma. J Pediatr (Rio J). February 2020; 97, pages37-43.
110. Mason Uvodich, Ross Barman, Adam Reitz, Matthew Sexton. A 16-Year-Old Male with Thoracic Compression following Posterior Spinal Instrumentation and Fusion for Marfan-Associated Syndromic Scoliosis. Hindawi Case Reports in Orthopedics.2020; Volume 2020, Article ID 6617028,5 pages1-5. https://doi.org/10.1155/2020/6617028.
111. Beop-Yong Lim, Youngwoong Kim, Hoseok I, Chi-Seung Lee. Numerical investigation of the sternoclavicular joint modeling technique for improving the surgical treatment of pectus excavatum. NATURERESEARCH. April 2020; ScientificRepots,10:7357.
112. Mayke J van Dort, Johanna HM Driessen, Elisabeth APM Romme, Piet Geusens, Paul C Willems, et la. Thoracic Kyphosis on Chest CT Scans Is Associated with Incident Vertebral Fractures in Smokers. JBMR. January 2019; Vol. 34, No. xx, page1-8. DOI: 10.1002/jbmr.3672.
113. Malay Sarkar, Rajeev Bhardwaz1, Irappa Madabhavi, Mitul Modi. Physical signs in patients with chronic obstructive pulmonary disease. Lung India. January-February 2019; Volume 36, Issue 1, pages 38-47.
114. Ravi S. Reddy, Khalid A. Alahmari, Paul S. Silvian, Irshad A. Ahmad, Venkata Nagaraj Kakarparthi et al. Reliability of Chest Wall Mobility and Its Correlation with Lung Functions in Healthy Nonsmokers, Healthy Smokers, and Patients with COPD. Canadian Respiratory Journal. February 2019; Article ID 5175949, pages1-11.

115. Kimia Khalatbari Kani, Hyojeong Mulcahy, Jack A. Porrino, Felix S. Chew. Thoracic cage injuries. EJRAD: ELSEVIER. 2019; vol.110, pages 225-232.

116. David Lo, Amit Maniyar, Sumit Gupta, Erol Gaillard. High prevalence of bronchiectasis on chest CT in a selected cohort of children with severe Asthma. BMC Pulmonary Medicine. 2019; 19:136, pages1-9. https://doi.org/10.1186/s12890-019-0900-0.

117. S J Lim, Ju-Young Kim, Seung Jun Lee, Gi Dong Lee, Yu Ji Cho, Yi Yeong Jeong et al. Altered Thoracic Cage Dimensions in Patients with Chronic Obstructive Pulmonary Disease Tuberculosis and Respiratory Diseases. 2018; Volume 81, pages 123-131.

118. Robert S Esinger, Hans K Pilegaard. Moving beyond the Haller Index as a measure of pectus excavatum severity. BJSTR: 2018; Florida, Vol.2, i.1, p 2129-2130.

119. Samuel Y. Ash, Alejandro A. Diaz. The Role of Imaging in the Assessment of Severe Asthma. Curr Opin Pulm Med. January 2017; 23(1), pages97–102. doi:10.1097/MCP.0000000000000341.

120. Siti W. Mohd-Zin, Ahmed I. Marwan, Mohamad K. Abou Chaar, Azlina Ahmad-Annuar, Noraishah M. Abdul-Aziz. Spina Bifida: Pathogenesis, Mechanisms, and Genes in Mice and Humans. Hindawi Scientifica. February2017; pages1-29. https://doi.org/10.1155/2017/5364827.

121. Emily M. DeBoer, David R. Spielberg, Alan S. Brody. Clinical potential for imaging in patients with asthma and other lung disorders. J Allergy Clin Immunol: CrossMark. January 2017; Volume 139, Number 1, pages 21-28.

122. Irfaan Abid, MennatAllah M. Ewais, Joseph Marranca, Dawn E. Jaroszewski. Pectus Excavatum: A Review of Diagnosis and Current Treatment Options. J Am Osteopath Assoc. February 2017; Vol 117 |No. 2, pages 106-113.

123. S.Singh, S.B. Greenberg. Hyperinflation of the left anterior basal segment: a new sign of pectus excavatum severity on CT. Clinical Radiology: ELSEVIER. June 2017; 72, 991.e15e-991.e18.

124. Anda Maria Aignătoaei, Cristina Elena Moldoveanu, Irina Draga Căruntu, Simona Eliza Giuşcă, Simona Partene Vicoleanu, Alin Horaţiu Nedelcu.

Incidental imaging findings of congenital rib abnormalities – a case series and review of developmental concepts. Folia Morphologica. August 2017; pages 1-17. ISSN:0015-5659; e-ISSN:1644-3284.

125. Nivedha Viswanathan,Ayesha Parveen, Yuvaraj Maria Francis, Priya,Dharshini gowthaman. A. Fusion of typical thoracic vertebrae. Int J Anat Res. 2017; 5(2.3):3923-26.

126. Prabu Rajkumar, Kamaraj Pattabi, Selvaraj Vadivoo, Arvind Bhome, Bill Brashier, Prashanta Bhattacharya et al. A cross-sectional study on prevalence of chronic obstructive pulmonary disease (COPD) in India: rationale and methods. BMJ Open. May 2017; 7:e015211. doi:10.1136/bmjopen-2016-015211.

127. Robert E. Kelly Jr, Robert J. Obermeyer, Donald Nuss. Diminished pulmonary function in pectus excavatum: from denying the problem to finding the mechanism. Annals of cardiothoracic surgery. September 2016; Vol 5, No 5, pages 466-475.

128. Guglielmo M, Riccardo I, Eliseo P, Salvatore M, Adolfo D, Giorgia R et. al. Anatomical variation: T1 spina bifida occulta. Radiological findings. ELSEVIER. December 2016; Radiology Case Reports 12 (2017) 207 – 209.

129. Fizan Abdullah, Jamie Harris. Pectus Excavatum: More Than a Matter of Aesthetics. Pediatric Annals. 2016; Vol. 45, No. 11, pages e403- 406.

130. Mak SM, Bhaludin BN, Naaseri S, Di Chiara F, Jordan S, Padley S. Imaging of congenital chest wall deformities. Br J Radiol. February 2016;89: 20150595.

131. Jin Yong Jeong, Joong Hyun Ahn, Sang Yong Kim, Yoon Hong Chun, Kyungdo Han, Sung Bo Sim, Keon Hyon Jo. Pulmonary function before and after the Nuss procedure in adolescents with pectus excavatum: Correlation with morphological subtypes. JCTS. March 2015; vol.10, issue37, pages1-5.

132. Mohd Nazeer, Surender M Rao, Simmi Soni, Sreekanth T, Ravinder M, Ramakranthi T. Fusion of Typical Thoracic Vertebrae: A Case Report. Sch. J. App. Med. Sci.2015; 3(1A):24-28.

133. Meena D. An unusual case of rib anomalies in number and morphology with association of vertebral and sternum anomalies: Computed tomography diagnosis. West Afr J Radiol [serial online] 2015: 22:101-4.
134. Patrick M. Poston, Sonali S. Patel, Maheen Rajput, Nicholas O. Rossi, Mohammed S. Ghanamah, James E. Davis et al. The Correction Index: Setting the Standard for Recommending Operative Repair of Pectus Excavatum. Annals of Thoracic Surg.2014; 97: pages1176–80.
135. Selvi Asker, Muntecep Asker, Bulent Ozbay. Evaluation of Airway Wall Thickness Via High-Resolution Computed Tomography in Mild Intermittent Asthma. Respiratory Care. April 2014; Volume 59, No 4, pages 550-555.
136. Dhana Lakshmi V, Arun Kumar K. R, Satish Kumar S, Suresh Kumar T, Suba Ananthi K.Multiple Bifid Ribs A Case Report. JEMDS: 2014; Vol. 3, Issue 15, p 3883-3886.
137. Nicola Sverzellati, Davide Colombi, Giorgia Randi, Antonio Pavarani, Mario Silva, Simon L. Walsh et a. Computed Tomography Measurement of Rib Cage Morphometry in Emphysema. PLOS ONE. July 2013; Volume 8, Issue 7, pages1-7, e68546.
138. Markus Bastir, Daniel García Martínez, Wolfgang Recheis, Alon Barash, Michael Coquerelle, Luis Rios et al. Differential Growth and Development of the Upper and Lower Human Thorax. PLOS ONE. September 2013; Volume 8, Issue 9, pages 1-13, e75128.
139. Michele Gaeta, Fabio Minutoli, Giuseppe Girbino, Alessandra Murabito, Caterina Benedetto, Rosario Contiguglia et al. Expiratory CT scan in patients with normal inspiratory CT scan: a finding of obliterative bronchiolitis and other causes of bronchiolar obstruction. Multidisciplinary Respiratory Medicine.2013; 8:44, pages1-8.
140. Rita Priori, Andrea Aliverti, André L. Albuquerque, Marco Quaranta, Paul Albert, Peter M. A. Calverley. The effect of posture on asynchronous chest wall movement in COPD. J Appl Physio. February2013; l114: pages1066–1075.

141. Mousa K, Jayakrishnan B, Abdulaziz M, Omolara R, Tariq S, Nabil M et al. High Resolution Computed Tomography in Asthma. Oman Medical Journal. January 2012; Vol. 27, No. 2, pages 145-150.

142. Ti-Hei Wu, Tsai-Wang Huang, Hsian-Her Hsu, Shih-Chun Lee, Ching Tzao, Huang Chan, Yeung-Leung Cheng. Usefulness of chest images for the assessment of pectus excavatum before and after a Nuss repair in adults. EJCTS. 15 May 2012; 43 (2013) 283–287.

143. Maree T Izatt, Clayton J Adam, Eugene J Verzin, Robert D Labrom, Geoffrey N Askin. CT and radiographic analysis of sagittal profile changes following thoracoscopic anterior scoliosis surgery. BioMed Central. 2012; Volume7, No15, pages1-13. http://www.scoliosisjournal.com/content/7/1/15.

144. M. Louise Lawson, Robert B. Mellins, James F. Paulson, Robert C. Shamberger, Keith Oldham, Richard G. Azizkhan et al. Increasing Severity of Pectus Excavatum is Associated with Reduced Pulmonary Function. JPEDS. August 2011; Vol. 159, No. 2, pages256-261

145. Wojciech G, Robert Sitnik MW, Hanna Kocon`, Pawel Bolewicki, Andrzej Górecki. Method of pectus excavatum measurement based on structured light technique. Journal of Biomedical Optics. August 2009; Vol. 14, issue4, pages 044041-8.

146. Michel M, Mathieu L, Ste´phanie R, Hugues B, Jose´ MT-de-Lara, Roger Marthan et al. Bronchial Measurements in Patients with Asthma: Comparison of Quantitative Thin -Section CT Findings with Those in Healthy Subjects and Correlation with Pathologic Findings. Radiology. December 2009; Volume 253, Number 3, pages 844-853.

147. A. Aliverti, M. Quaranta, B. Chakrabarti, A.L.P. Albuquerque, P.M. Calverley. Paradoxical movement of the lower ribcage at rest and during exercise in COPD patients. Eur Respir J. 2009; Vol 33, Issue1, pages 49–60.DOI: 10.1183/09031936.00141607.

148. Fokin AA, Steuerwald NM, Ahrens WA, Allen KE. Anatomical, histologic, and genetic characteristics of congenital chest wall deformities. Semin

Thorac Cardiovasc Surg 2009; 21:44–57. doi: http://dx.doi.org/10.1053/j.semtcvs.2009.03.001.

149. Marcus P. Kennedy, Peadar G. Noone, Margaret W. Leigh, Maimoona A. Zariwala, Susan L. Minnix, Michael R. Knowles et al. High-Resolution CT of Patients with Primary Ciliary Dyskinesia. AJR. May 2007; 188:1232–1238.0361–803X/07/1885–1232.

150. Dawn E. Jaroszewski, Eric W. Fonkalsrud. Repair of Pectus Chest Deformities in 320 Adult Patients: 21 Year Experience. Ann Thorac Surg. 2007; 84:429 –33.

151. Deborah M. Kado, Katherine Prenovost, Carolyn Crandall. Narrative Review: Hyperkyphosis in Older Persons. Annals of Internal Medicine. September 2007; Volume 147, Number 5, pages 330-338.

152. P.A. de Jong, N.L.Mu¨ller, P.D. Pare, H.O. Coxson. Computed tomographic imaging of the airways: relationship to structure and function. Eur Respir J. March 2005; Volume 26, Number 1, pages 140–152.

153. V. Ne`ve, F. Girard, A. Flahault, M. Boule. Lung and thorax development during adolescence: relationship with pubertal status. Eur Respir J 2002; 20: 1292–1298. DOI: 10.1183/09031936.02.00208102.

154. Ronald B. J. Glass, Karen I. Norton, Sandra A. Mitre, Eugene Kang. Pediatric Ribs:A Spectrum of Abnormalities. RadioGraphics. January-February2002; Volume 22, No1, pages87-104.

155. Adam R. Guttentag, Julia K. Salwen. Keep Your Eyes on the Ribs: The Spectrum of Normal Variants and Diseases That Involve the Ribs. Department of Radiology, Albert Einstein Medical Center, Old York Rd, Philadelphia. January1999; 19: pages1125–1142.

156. David S. Gierada, Shahin Hakimian, Richard M. SIone, Roger D. Yusen. MR Analysis of Lung Volume and Thoracic Dimensions in Patients with Emphysema Before and After Lung Volume Reduction Surgery. AJR. March 1998; 170, pages 707-714.

157. Donald Nuss, Robert E. Kellyr, Jr. Daniel P. Croitoru, Michael E. Katz, Norfolk, Virginia. A lo-Year Review of a Minimally Invasive Technique for

the Correction of Pectus Excavatum. Journal of Pediatric Surgery. April 1998; Volume 33, No 4, pages.545-552.

158. Janet E. Kuhlman, Benedicto L. Reyes, Ralph H. Hruban, Frederic B. Askin, Elias A. Zerhouni, Elliot K Fishman et al. Abnormal Air-filled Spaces in the Lung. Radiographics 1993; vol.13: pages.47-75.

159. Haller JA Jr, Kramer SS, Lietman SA. Use of CT scans in selection of patients for pectus excavatum surgery: a preliminary report. J Pediatric Surg. 1987; 22:904–6.doi: http://dx.doi.org/10.1016/S0022-3468(87) 80585-7. (78 ref in introduction)

160. Gerald T,Fon, Michael J Pitt, A Cole Thies, Jr. Thoracic Kyphosis: Range in Normal Subjects. AJR. May 1980; Volume 134, pages 979-983.

161. J. L. Ehrenhaft, Nicholas P. Rossi, M. S. Lawrence. Developmental Chest Wall Defects. THE ANNALS OF THORACIC SURGERY. May 1966; vol2, No3, pages384-98.

162. Suzanne L. Aquino, Gregory R. Duncan, L. Anne Hayman et. al. Nerves of the Thorax: Atlas of Normal and Pathologic Findings. Radio Graphics September-October 2001; vol21, No5, pages1275–1281.

Methodology -II (HRCT Criteria)

163. Yamagishi M. Computed tomographic-pathologic correlation of centriacinar emphysema. Nippon Kyobu Shikkan Gakkai Zasshi 1990; 28:681-690(in Japanese).

164. Peat JK, Woolcock AJ, Cullen K. Rate of decline of lung function in patients with asthma. Eur J Respir Dis1987; 70:171–9.

165. Ulrik CS, Backer V, Dirksen A. A ten-year follow-up of 180 adults with bronchial asthma: factors important for the decline of lung function. Thorax1992; 47:14–8.

166. Knox AJ. Airway remodelling in asthma: role of airway smooth muscle. Clin Sci1994; 86:647–52.

167. Pare PD, Bai TR. The consequences of chronic allergic inflammation. Thorax 1995; 50:328–32.

168. Redington AE, Howarth PH. Airway remodeling in asthma. Thorax1997; 52:310–2.
169. Hogg JC, Macklem PT, Thurlbeck WM. Site and nature of airway obstruction in chronic obstructive lung disease.N Engl J Med1968;278:1355–60.
170. Foster WL, Pratt P, Roggli V, et al. Centrilobular emphysema/CT-pathologic correlation. Radiology 1986; 159:27–32.
171. Muller NL, Miller RR, Abboud RT. "Density mask". An objective method to quantify emphysema using computed tomography. Chest1988; 94:782–7.
172. Kuwano K, Matsuba K, Ikeda T, et al. The diagnosis of mild emphysema. Correlation of computed tomography and pathology scores. Am Rev Respir Dis1990; 141:169–78.
173. Bergin C, Muller N, Nichols DM, et al. The diagnosis of emphysema. A computed tomographic-pathologic correlation. Am Rev Respir Dis1986; 133:541–6.
174. Goddard PR, Nicholson EM, Laszlo G, et al. Computed tomography in pulmonary emphysema. Clin Radiol1982; 33:379–87.
175. Nestor L. Muller. CT Diagnosis of Emphysema, it may be accurate, but is it relevant? CHEST. February1993; Columbia, Vol.103, Issue 2, page 329-330.
176. Thurlbeck WM. Morphology of emphysema and emphysema-like conditions. Philadelphia: Saunders, 1976:9.
177. Orlandi I, Moroni C, Camiciottoli G, Bartolucci M, Pistolesi M, Villari N, Mascalchi M. Chronic obstructive pulmonary disease: thin-section CT measurement of airway wall thickness and lung attenuation. Radiology2005,234:604–610.
178. Jeong Uk lim, Ji-hyun lee, Tae-hyung Kim, Jae seung lee, Sang-Do lee, Yeon-Mok Oh, Chin Kook Rhee. Alternative definitions of chronic bronchitis and their correlation with CT parameters. International Journal of COPD. 2018:13 1893–1899.

179. Hansell DM, Bankier AA, MacMahon H, McLoud TC, Müller NL, Remy J. Fleischner Society: glossary of terms for thoracic imaging. Radiology 2008;246(3):697–722.

180. Bethany Milliron, Travis S. Henry, Srihari Veeraraghavan, Brent P. Little. Bronchiectasis: Mechanisms and Imaging Clues of Associated Common and Uncommon Diseases. Radiographics, July-August 2015; Volume35, Number4, page 1011-1029.

181. Smith IE, Jurriaans E, Diederich S, Ali N, Shneerson JM, Flower CD. Chronic sputum production: Correlations between clinical features and findings on high resolution computed tomographic scanning of the chest. Thorax 1996; 51:914-8.

182. Kim JS, Müller NL, Park CS, Grenier P, Herold CJ. Cylindrical bronchiectasis: Diagnostic findings on thin-section CT. AJR Am J Roentgenol 1997; 168:751-4.

183. Marti-Bonmati L, Catala FJ, Ruiz Perales F. Computed tomography differentiation between cystic bronchiectasis and bullae. J Thorac Imaging 1991; 7:83-5.

184. Abdulaziz H. Alzeer. HRCT score in bronchiectasis: Correlation with pulmonary function tests and pulmonary artery pressure. Annals of Thoracic Medicine. July-September 2008; Vol 3, Issue 3, page 82-86.

185. Travis WE, Colby TV, Koss MN, Rosado de Christenson ML, Müller NL.Non-neoplastic disorders of the lower respiratory tract. Washington, DC: Armed Forces Institute of Pathology, 2002:381–471.

186. C Isabela S. Silva, Thomas V. Colby, Nestor L. Müller. Asthma and Associated Conditions:

187. High-Resolution CT and Pathologic Findings. AJR January 30, 2004. Canada: 183:817–824 0361–803X/04/1833–817.

188. Lynch DA, Newell JD, Tschomper BA, Cink TM, Newman LS, Bethel R. Uncomplicated asthma in adults: comparison of CT appearance of the lungs in asthmatic and healthy subjects. Radiology1993; 188:829–833.

189. Grenier P, Mourey-Gerosa I, Benali K, et al. Abnormalities of the airways and lung parenchyma in asthmatics: CT observations in 50 patients and inter- and intraobserver variability. Eur Radiol 1996; 6:199–206.

190. Park CS, Müller NL, Worthy SA, Kim JS, Awadh N, Fitzgerald M. Airway obstruction in asthmatic and healthy individuals: inspiratory and expiratory thin-section CT findings. Radiology1997; 203:361–367.

191. Aas K. Heterogeneity of bronchial asthma. Allergy1981; 36:3–10.

192. Henary Gray edited by Susan Standring. The Anatomical Basis of Clinical Practice. Elsevier limited; 2016, 41 editions, Chap 51, p. 545-46.

193. E. Kupeli, G.Ulubay. Bony Bridge of a bifid rib. Cleveland Clinic Journal of Medicine, vol 77, pp. 4232-233, 2010.

194. J. A. Geel, K. G. Bennett, J. M. Rigby, J. E. Poole. Gorlin syndrome. SA Journal of Child Health, vol 5, no.1, pp.21-22, Mar, 2011.

195. Pilegaard H (2015) Nuss technique in pectus excavatum: a monoinstitutional experience.J Thorac Dis7(Suppl 2), S172–S176.

196. RAVITCH MM. Unusual sternal deformity with cardiac symptoms operative correction. J Thorac Surg. 1952 Feb;23(2):138-44.

197. Kálmán A. Initial results with minimally invasive repair of pectus carinatum. J Thorac Cardiovasc Surg. 2009 Aug;138(2):434-8.

198. Hong J-Y, Suh S-W, Easwar T, et al. (2011) Evaluation of the three-dimensional deformities in scoliosis surgery with computed tomography: efficacy and relationship with clinical outcomes. Spine 36, E1259–E1265.

199. Albertal M, Vallejos J, Bellia G, et al. (2013) Changes in chest compression indexes with breathing underestimate surgical candidacy in patients with pectus excavatum: a compute tomography pilot study. J. Pediatr Surg48, 2011–2016.

200. Birkemeier K, Podberesky D, Salisbury S, et al. (2011) Breathe in... breathe out...stop breathing: does phase of respiration affect the Haller index in patients with pectus excavatum? AJR Am J Roentgenol197, W934–W939.

201. Kasper DL, Fauci AS, Hauser S, et al, editors. Harrison`s Principle of Internal Medicine. Mc Graw Hill Education; 2018, 20 editions, pages1997.

202. Sverzellati N, Colombi D, Randi G, Pavarani A, Silva M, Walsh SL, et al. Computed tomography measurement of rib cage morphometry in emphysema. PLoS One 2013;8: e68546.
203. Cassart M, Gevenois PA, Estenne M. Rib cage dimensions in hyperinflated patients with severe chronic obstructive pulmonary disease. Am J Respir Crit Care Med 1996;154(3 Pt 1):800-5.
204. Arakawa H, Kurihara Y, Nakajima Y, Niimi H, Ishikawa T, Tokuda M. Computed tomography measurements of overinflation in chronic obstructive pulmonary disease: evaluation of various radiographic signs. J Thorac Imaging 1998; 13:188- 92.
205. Bartynski WS, Heller MT, Grahovac SZ, Rothfus WE, Kurs Lasky M. Severe thoracic kyphosis in the older patient in the absence of vertebral fracture: association of extreme curve with age. AJNR Am J Neuroradiol 2005; 26:2077-85.
206. Sharma G, Goodwin J. Effect of aging on respiratory system physiology and immunology. Clin Interv Aging 2006; 1:253- 60.
207. Bellemare F, Jeanneret A, Couture J. Sex differences in thoracic dimensions and configuration. Am J Respir Crit Care Med 2003; 168:305-12.
208. Weaver AA, Schoell SL, Stitzel JD. Morphometric analysis of variation in the ribs with age and sex. J Anat 2014; 225:246-61.
209. Salito C, Luoni E, Aliverti A. Alterations of diaphragm and rib cage morphometry in severe COPD patients by CT analysis. Conf Proc IEEE Eng Med Biol Soc 2015; 2015:6390-3
210. Lando Y, Boiselle P, Shade D, Travaline JM, Furukawa S, Criner GJ. Effect of lung volume reduction surgery on bony thorax configuration in severe COPD. Chest 1999; 116:30-9.
211. Eberlein M, Reed RM, Maidaa M, Bolukbas S, Arnaoutakis GJ, Orens JB, Brower RG, Merlo CA, Hunsicker LG. Donor-recipient size matching and survival after lung transplantation. A cohort studies. Ann Am Thorac Soc 2013; 10: 418-425 [PMID: 23988005 DOI: 10.1513/AnnalsATS.201301-008OC].

212. Eberlein M, Diehl E, Bolukbas S, Merlo CA, Reed RM. An oversized allograft is associated with improved survival after lung transplantation for idiopathic pulmonary arterial hypertension. J Heart Lung Transplant 2013; 32: 1172-1178 [PMID: 23876630 DOI: 10.1016/j.healun.2013.06.011]
213. Eberlein M, Reed RM, Permutt S, Chahla MF, Bolukbas S, Nathan SD, Iacono A, Pearse DB, Fessler HE, Shah AS, Orens JB, Brower RG. Parameters of donor-recipient size mismatch and survival after bilateral lung transplantation. J Heart Lung Transplant 2012; 31: 1207-1213.e7 [PMID: 22036314 DOI: 10.1016/ j. healun.2011.07.015]
214. McNulty W, Jordan S, Hopkinson NS. Attitudes and access to lung volume reduction surgery for COPD: a survey by the British Thoracic Society. BMJ Open Respir Res 2014; 1: e000023.Abstract/FREE Full TextGoogle Scholar
215. Fishman A, Martinez F, Naunheim K. A randomized trial comparing lung-volume-reduction surgery with medical therapy for severe emphysema. N Engl J Med 2003; 348: 2059–2073.CrossRefPubMedWeb of ScienceGoogle Scholar
216. Deslee G, Klooster K, Hetzel M, et al. Lung volume reduction coil treatment for patients with severe emphysema: a European multicentre trial. Thorax 2014; 69: 980–986.Abstract/FREE Full TextGoogle Scholar
217. Sciurba FC, Ernst A, Herth FJF, et al. A randomized study of endobronchial valves for advanced emphysema. N Engl J Med 2010; 363: 1233–1244.CrossRefPubMedWeb of ScienceGoogle Scholar
218. Palla A. Elective surgery for giant bullous emphysema: a 5-year clinical and functional follow-up. Chest J 2005; 128: 2043.
219. Pauwels RA, Buist AS. 2001 Global strategy for the diagnosis, management, and prevention of chronic obstructive pulmonary disease. NHLBI/WHO Global Initiative for Chronic Obstructive Lung Disease (GOLD) Workshop summary. Am J Respir Crit Care Med 163 1256 76.
220. Kurashima K. Takayanagi N. Sato N. et al. 2005 High resolution CT and bronchial reversibility test for diagnosing COPD. Respirology 2005; 10 316 22.

221. Morgan MDL. 1992 Detection and quantification of pulmonary emphysema by computed tomography: a window of opportunity. Thorax 47 10014.
222. Rosenblum LJ, Mauceri RA, Wellenstein DE, et al. Computed tomography of the lung. Radiology 1978; 129: 521–524.
223. Patel B. Make B. Coxson H. O. et al. 2006 Airway and parenchymal disease in chronic obstructive pulmonary disease are distinct phenotypes. Proc Am Thorac Soc 3: 533.
224. Boschetto P. Miniati M. Miotto D. et al. 2003 Predominant emphysema phenotype in chronic obstructive pulmonary. Eur Respir J 21 450 4.
225. Boschetto P. Quintavalle S. Zeni E. et al. 2006 Association between markers of emphysema and more severe chronic obstructive pulmonary disease. Thorax 61 1037 42.
226. Grenier P. A. Beigelman-Aubry C. Fetita C. et al. 2004 Large airways at CT: Bronchiectasis, Asthma and COPD. In: Kauczor HU, editor. Functional imaging of the chest. Heidelberg: Springer; 39 55.
227. Martinez F. J. Foster G. Curtis J. L. et al. 2006 Predictors of mortality in patients with emphysema and severe airflow obstruction. Am J Respir Crit Care Med 173 1326 34.
228. Aziz ZA, Wells AU, Desai SR, et al. 2005 Functional impairment in emphysema: contribution of airway abnormalities and distribution of parenchymal disease. Am J Roentgenol 185 1509 15.
229. Nakano Y. Sakai H. Muro S. et al. 1999 Comparison of low attenuation areas on computed tomographic scans between inner and outer segments of the lung in patients with chronic obstructive pulmonary disease: incidence and contribution to lung function. Thorax 54 384 9.
230. Gould G. A. Mac Nee. W. Mclean A. 1988 CT measurements of lung density in life can quantitate distal airspace enlargement: an essential defining feature of human emphysema. Am Rev Resp Dis 133 380 92.
231. Sanders C. Nath P. H. Bailey W. C. 1988 Detection of emphysema with computed tomography: correlation with pulmonary function tests and chest radiography. Invest Radio 23 262 6.

232. Gupta P,P. Yadav R. Verma M. Agarwal D. Kumar M.et. al.2008 Correlation between high resolution computed tomography features and patients` characteristics in chronic obstructive pulmonary disease. Annals Thorac Med 3 87 93.

233. Gupta P.. P. Yadav, R. Verma M., Gupta K., B. Agarwal D et. al. 2009 High resolution computed tomography features in patients with chronic obstructive pulmonary disease. Singapore Med J 50 193 200.

234. Nowell Jr JD. CT of emphysema. 2002 Radiologic Clinics of North America 40 1 31 42.

235. Harrison G A, Weiner J S, Tanner J M, Barnicot NA. Human biology. Clarendon press: Oxford, 1964, chap 19.

236. Biavati M, Kozlitina J, Alder AC, Foglia R, McColl RW, Peshock RM, et al. (2020) Prevalence of pectus excavatum in an adult population-based cohort estimated from radiographic indices of chest wall shape. PLoS ONE 15(5): e0232575.

237. Westphal FL, Lima LC, Lima Neto JC, Chaves AR, Santos Junior VL, Ferreira BL. Prevalence of pectus carinatum and pectus excavatum in students in the city of Manaus, Brazil. J Bras Pneumol. 2009; 35(3):221–6. Epub 2009/04/25. https://doi.org/10.1590/s1806-37132009000300005 PMID: 19390719.

238. Coskun ZK, Turgut HB, Demirsoy S, Cansu A. The prevalence and effects of Pectus Excavatum and Pectus Carinatum on the respiratory function in children between 7–14 years old. Indian J Pediatr.2010; 77(9):1017–9. Epub 2010/09/04. https://doi.org/10.1007/s12098-010-0155-5 PMID: 20814837.

239. Cartoski MJ, Nuss D, Goretsky MJ, Proud VK, Croitoru DP, Gustin T, et al. Classification of the dysmor-phology of pectus excavatum. Journal of pediatric surgery. 2006; 41(9) : 1573– 81. https: //doi. org/ 10. 1016/ j. jpedsurg.2006.05.055 PMID: 16952594.

240. Prasad, H., Parashar, R., & Jain, A. K. (2022). Changes in thoracic form morphology in healthy young adults to middle-aged people at numerous

thoracic vertebral levels. International Journal of Health Sciences, 6(S2), 11514–11524.

241. S. Allwyn Joshua, Lathika Shetty, Vidyashambhava Pare, Sheela G Nayak and Roopa Sebastian (2014), "Radiological Indices for Chest in the Indian Population", Journal of Radiology Research and Practice, Vol. 2014 (2014), Article ID 145020, DOI: 10.5171/2014.145020

242. Edward Cantu, Andrew Courtwright et.al. Evaluation and Management ofthe Potential Lung Donor Clin Chest Med. 2017 December ; 38(4): 751–759. doi:10.1016/j.ccm.2017.07.007.

243. Mallory GB Jr. Donor-recipient lung size mismatch: a provocative data set. J Heart Lung Transplant. 2013; 32:1163–4.

244. Reitz BA, Wallwork JL, Hunt SA, et al. Heart-lung transplantation: successful therapy for patients with pulmonary vascular disease. N Engl J Med. 1982; 306:557–64. https://doi.org/10.1056/NEJM198203113061001.

245. Goldstein R, Bolman R, Cooper J, et al. Lung transplantation: report of the ATS Workshop on Lung Transplantation. Am Rev Respir Dis. 1993; 147:772–6.

246. Mason DP, Batizy LH, Wu J et al. Matching donor to recipient in lung transplantation: How much does size matter? J Thorac Cardiovasc Surg. 2009; 137:1234–40 e1.

247. Chacon RA, Corris PA, Dark JH, Gibson GJ. Respiratory mechanics after heart-lung and bilateral lung transplantation. *Thorax.* (1997) 52:718–22. doi: 10.1136/thx.52.8.718.

248. Geddes D, Davies M, Koyama H, Hansell D, Pastorino U, Pepper J, et al. Effect of lung-volume-reduction surgery in patients with severe emphysema. *N Engl J Med.* (2000) 343: 239–45. Doi: 10. 1056/ nejm 200007273430402

249. Fishman A, Martinez F, Naunheim K, Piantadosi S, Wise R, Ries A, et al. A randomized trial comparing lung-volume-reduction surgery with medical therapy for severe emphysema. *N Engl J Med.* (2003) 348:2059–73. doi: 10.1056/NEJMoa030287.

250. Gilmartin JJ, Gibson GJ. Abnormalities of chest wall motion in patients with chronic airflow obstruction. *Thorax.* (1984) 39:264–71. doi: 10.1136/thx.39.4.264.

251. Zoumot Z, LoMauro A, Aliverti A, Nelson C, Ward S, Jordan S, et al. Lung volume reduction in emphysema improves chest wall asynchrony. *Chest.* (2015) 148:185–95. doi: 10.1378/chest.14-2380

252. Jenkins SC, Moxham J. The effects of mild obesity on lung function. *Respir Med.* (1991) 85:309–11. doi: 10.1016/s0954-6111(06)80102-2.

253. Boriek AM, Lopez MA, Velasco C, Bakir AA, Frolov A, Wynd S, et al. Obesity modulates diaphragm curvature in subjects with and without COPD. *Am J Physiol Regul Integr Comp Physiol.* (2017) 313: R620–9. doi: 10.1152/ajpregu.00173.2017

254. Bellemare JF, Cordeau MP, Leblanc P, Bellemare F. Thoracic dimensions at maximum lung inflation in normal subjects and in patients with obstructive and restrictive lung diseases. *Chest.* (2001) 119:376–86. doi: 10.1378/chest.119.2.376.

www.ingramcontent.com/pod-product-compliance
Lightning Source LLC
LaVergne TN
LVHW080050210926
848481LV00011B/3320

* 9 7 9 8 2 3 0 4 7 4 6 5 4 *